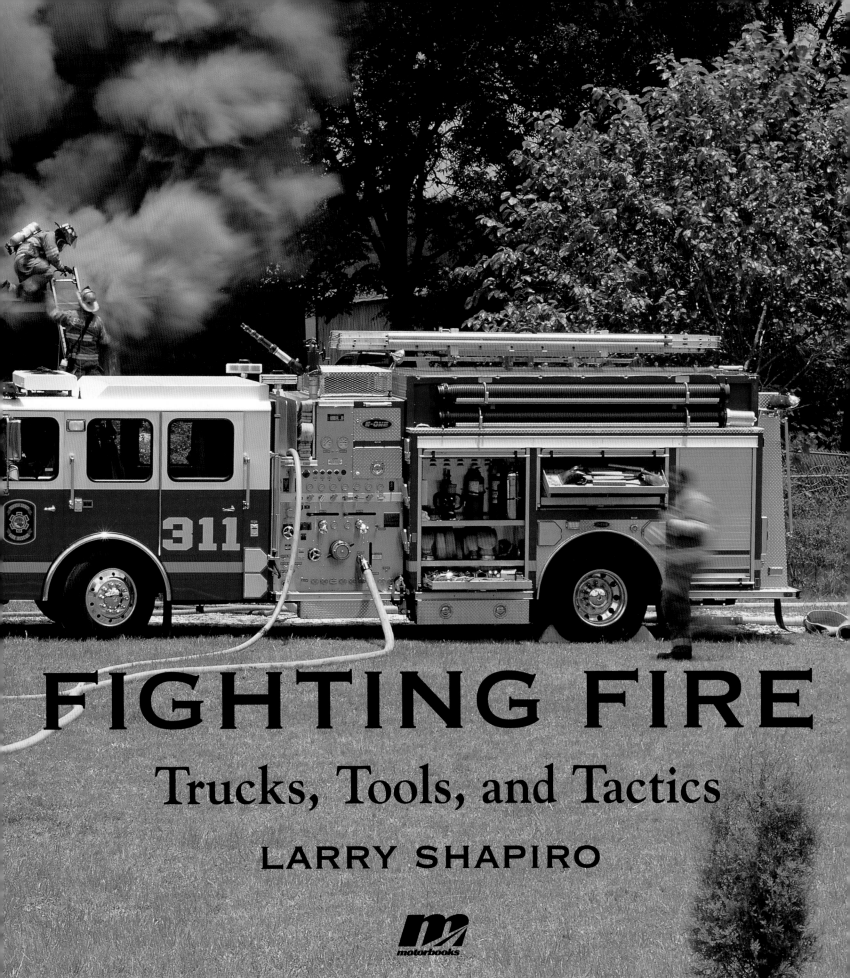

FIGHTING FIRE

Trucks, Tools, and Tactics

LARRY SHAPIRO

motorbooks

Editor: Peter Schletty
Designer: Chris Fayers

Printed in China

On the cover: Chicago engine 126 has several lines off at this 3–11 alarm fire on the city's south side. Burning out of control is a supermarket. Farther down the street, one of several aerials can be seen through the smoke deploying an elevated master stream. Chicago runs 99 engine companies. The vast majority of the engines were built by Luverne or Crimson, which is the new name for the company that combined Luverne with Quality Manufacturing in 2003. Although the early models that were sold to the city were built on HME chassis, the bulk feature Spartan chassis made by Spartan Motors, the parent company of Crimson. *Steve Redick*

On the frontispiece: Firefighters drag hose through eight inches of water that has accumulated in the parking lot of a lumberyard after the fire has been knocked down. This hose will extend a line that is already in place, allowing firefighters to venture closer to the building to extinguish the remaining hot spots.

On the title pages: As firefighters put a line into the front door, the roof firefighter who was venting the attic window is guided through the dense smoke to the ladder so he can climb down. Engine 1711 is nursing engine 311 at this house fire in the country prior to the arrival of several water tankers/tenders for a sustained water supply. The rigs are twin E-ONE custom Cyclone pumpers.

On the back cover, top: Industrial firefighting is often about applying enormous volumes of water or foam for a prolonged period of time to extinguish, protect, or cool elements in the plant. This is why industrial rigs have built-in foam proportioning systems and big deluge guns capable of discharging up to 4,000 gallons of water per minute. **Bottom:** Firefighters most often use water as the primary extinguishing agent, but some departments prefer using foam. Certain foams have the ability to coat the walls and other surfaces around a fire, robbing the fire of burnable materials.

TABLE OF CONTENTS

INTRODUCTION

Firefighting has always been about more than simply dousing fires. It's about saving and protecting lives and property, being there to help the public, and performing unselfish acts.

Over the years, the tasks that have been asked of firefighters have grown in number, in kind, in severity, and in frequency. Today's fire departments continue to fight fires, perform rescues, and respond to motor-vehicle accidents and calls for emergency medical services (EMS). The men and women of the fire service also respond to natural disasters including hurricanes, flooding, snowstorms, earthquakes, and forest fires, and to calls for international aid when disasters strike across the globe. Firefighters manage industrial accidents, building collapses, chemical spills, and bombings. Firefighters today are qualified to handle hazardous-materials incidents, terrorist threats, and large-scale incidents involving mass casualties. Firefighters are a major part of our nation's first responders, whether they work in a major city, a suburban fire department, a rural area, or an industrial setting.

All firefighters share common skills inherent to the job. At the same time, the diverse environments in which they work present unique challenges.

Urban firefighters are exposed to different obstacles than firefighters in many suburban departments, while rural fire departments have their own set of hurdles to overcome in the performance of their duties. The intricacies of attacking a forest fire differ greatly from the complexities of structural firefighting in the dead of winter, while chemical fires within large industrial plants require tactics and procedures that have little in common with fires in remote rural areas.

This book will illustrate similarities and differences that are inherent in urban, suburban, rural, industrial, and wildland firefighting. The manpower, the trucks, the equipment used, and the tactics involved in each type of firefighting will be examined.

In addition, this book will highlight other aspects of being a firefighter: the tools and apparatus they utilize and the growing resources devoted to EMS and specialized rescue operations.

No two fire departments are alike. This book will attempt to give an overview of all types of departments, but it by no means has the ability to be inclusive of the entire fire service. Therefore, apologies are extended to those fire departments that feel their particular situations have not been addressed.

ACKNOWLEDGMENTS

Writing a book of this nature requires research, some skill, patience, and the help of many outgoing individuals. Since I am not a firefighter, I had to rely on many sources in order to compile all of the information contained in this book. I have had the good fortune to spend several years working with various facets of the fire service and, as such, have garnered many contacts and, more importantly, friends. I reached out to the professionals acknowledged here, whom I've had the pleasure of working with or those whom I have never met but who were gracious enough to share their expertise with me through referrals. The names represent quite a bit of history, longevity, experience, and knowledge in all aspects of the fire service. I merely acted as the conduit to try to seam together some of their wisdom into a cohesive, informative, and entertaining book.

Alphabetically, my thanks and gratitude go out to: Bruce Boyle, Assistant Chief, Elwood Fire Protection District, IL; Patrick Butler, Battalion Chief, Los Angeles City Fire Department, CA; Mike Chandler, Fire Chief, Rumsey-Rancheria Fire Department, Brooks, CA; Ron Coleman, California State Fire Marshall, retired; Sean Evans, Captain, Fairfax County Fire and Rescue Department, VA; David

Opposite: There are different schools of thought on the advantages and disadvantages of live-fire training in an abandoned structure. Many find the value of the scenarios available superior to the controlled environments at specialized fire training facilities. One point that is not disputed is the critical need for highly qualified and experienced instructors to ensure the safety of all personnel in the donated buildings. Another aspect that everyone can agree on is that the resulting fires are spectacular.

Acknowledgments

Franklin, Captain, San Francisco Fire Department, CA; Don Frazeur, Assistant Chief, Los Angeles City Fire Department, CA; James S. Griffiths, Honorary Chief of Operations, FDNY, author, *Fire Department of New York: An Operational Reference*, 2007; Bill Gustin, Captain, Miami-Dade Fire Rescue, FL; Jack Hickey, Assistant Chief, Davis Fire Department, CA; Buddy Jackson, Eddie Bagley, James Mosley, Mike Meadows, and Benni Endel, Lion Oil Company Fire Brigade, El Dorado, AR; Jerry Keohane, Captain, San Francisco Fire Department, CA; Ken Koerber, Rescue Manager, IL US&R TF-1, Battalion Chief, Deerfield-Bannockburn Fire Protection District, IL; Jack Lerch, Honorary Chief of Department, FDNY; Frank Lucca, Lieutenant, Buffalo Fire Department, NY; John Martinez, Battalion Chief, Los Angeles City Fire Department, CA; George Mattos, Fire Equipment Manager I, CAL FIRE; David M. McGrail, District Chief, Denver Fire Department, CO, author, *Firefighting Operations in High-Rise and Standpipe Equipped Buildings*; Barry McRoy, Director, Colleton County Fire-Rescue, SC; Mikel Milks, Director, Glencoe Department of Public Safety, IL; Moon Mullens, retired Fire Chief, Exxon Mobil, Baton Rouge, LA; Scott G. Nacheman, Structures Specialist, DHS/FEMA US&R IN-TF1, IST; Field Instructor, Illinois Fire Service Institute; Frank Sandrock, Captain Rescue One, Camden Fire Department, NJ; Drew Smith, Assistant Chief, Prospect Heights Fire Protection District, IL; Kevin Story, Captain, Houston Fire Department, TX; Brad Targhetta, retired Fire Chief, Brighton Fire Department, IL; Tom Tarp, Division Chief of Operations retired, California Department of Forestry and Fire Protection; Senior Associate, Thornton Tomasetti, Chicago, IL; David Traiforos, Fire Chief, Franklin Park Fire Department, IL; Michael Ursitti, CAL FIRE Forestry Equipment Manager II, Northern Region; Tom Whittaker, Wheeling Fire Department, IL; Stephen Wilcox, retired Lieutenant, Wheaton Fire Department, IL, Fire Photographer, Dupage County, IL; and Tripp Wilson, Arlington Heights Fire Department, IL.

Although the majority of images in this book are mine, there are several that were submitted by talented photographers from across the country. Each image is credited to them as they appear throughout the book, but I want to introduce them here. My thanks go out to Jon Androwski, Jay K. Bradish, Michael J. Coppola, Keith Cullom, Matt Daly, Rick McClure, Ted Pendergast, Steve Redick, and Dan Reiland, for helping to enhance this book.

In addition, I want to thank fellow fire geeks: Scott Lasker, Tim Olk, Steve Redick, Dave Traiforos, Shaun Unell, Tom Whittaker, and Tripp Wilson, who all share the passion and remain in constant communication to make sure that some if not all can get to the fires and wrecks.

As I mentioned in the opening of these acknowledgments, one of the requirements involved with this book was patience. Certainly the only thing I had to be patient about was the fact that wherever I traveled there just were no fires. The patience that is meaningful to me, though, is that of my wife, Dorothy. She has to put up with me traveling, sitting for hours in front of the computer, listening constantly to fire radios, running out to incidents at all hours of the day and night looking for that shot that will enhance the book, and lastly, the patience to read and reread every word I've typed so that I convey some semblance of knowledge. My wife has been with me through several books and actually had the chance to utilize her fire service experience when she was hired for a short time to work for James Lee Witt and Associates while they conducted an inquiry into the circumstances of the fatal fire at the Cook County Administration Building in Chicago on October 17, 2003. Boy were they impressed that she knew all about a 3-11 and the CFD radio jargon!

Dorothy and I are blessed with good friends, family, and three wonderful boys. We are proud of them all but cannot help but maintain an added sense of fulfillment with our middle son, Paul Graf, who at the time of this writing is serving in the U.S. Army in Iraq. Our thoughts and prayers are with him and all of the men and women in the armed forces who are serving our country.

Opposite: Firefighters from Truck 16 in Skokie, Illinois, work to ventilate the roof of this single-family house. The fire started in a car parked near the house and quickly spread through the soffit into the attic. Firefighters have vented on the one end and continue to open the roof with a saw and an axe so the fire will work its way up through the roof instead of down into the house. Firefighters making an interior attack are taking a beating and need the vertical ventilation so they can advance on the fire in hopes of preventing its spread to the remainder of the house. *Steve Redick*

CHAPTER ONE

SUBURBAN FIREFIGHTING

Fire Apparatus

Fire departments have several types of apparatus, or rigs, to utilize in their fleets. The most common is the engine, or pumper. This rig has an integral fire pump, and a water tank, and it carries several types of hose for supply or attack, ground ladders, hand tools, fittings for hydrants and hose lines, and anywhere from two to six firefighters.

The next rig is a truck. Trucks carry more ground ladders than engines, have hand tools for forcible entry and rescue, and are staffed by as few as two firefighters or as many as six. Trucks do not carry water, although some departments will outfit a truck with a small water tank. A truck has a permanent aerial device that is capable of extending or elevating to the upper floors or the roof of a building. The aerials can range from extended lengths of 50 feet to 134 feet. The aerial can be a ladder, a tower ladder, or an elevating platform. The tower ladder has a platform or bucket that can support several firefighters at the end of the telescopic ladder. An elevating platform is a unit that articulates at one or more joints instead of telescoping out in a straight line.

Another rig is the quint, short for quintuple combination pumper. It is the combination of an engine with at least 300 gallons of water and a truck with a minimum amount of hose, equipment storage, and ground ladders. Quint staffing is comparable to an engine or a truck, depending on the department.

Fire departments vary in the terminology used for the rescue or squad company. This is a large toolbox on wheels. Traditionally, this unit does not have water and may not carry any ladders. The rescue or squad (or rescue squad) carries specialized tools for extrication and forcible entry, and may be outfitted for special types of rescues in the water, below the ground, or high above the ground. Many are also equipped for hazardous-materials incidents. Manpower for this company again can range from three to six firefighters, who often supplement the work of truck companies on the fire scene.

Areas that lack fire hydrants often require the fire department to carry larger quantities of water with them to the scene. These rigs are known as tankers or tenders. In some parts of the country, the term *tanker* refers to a plane, so the fire service is trying to standardize the term *tender* nationwide. Most often, staffing will call for fewer personnel on these rigs, since their primary function is to shuttle water to the scene to supply an engine and then leave to refill and return again.

Many fire departments provide emergency medical services (EMS). Firefighters trained to provide medical treatment often respond in units capable of patient transport. Commonly known as an ambulance, these may also be referred to as rescues, squads, medic units, or medic rescues. Two or three firefighters trained to the emergency medical technician (EMT) or paramedic level are common on these units. Department protocol may or may not include these individuals as fire suppression personnel on the scene of a fire. Some strictly perform medical duties.

Tasks

The firefighters assigned to each rig are referred to as a company. Firefighters wear turn-out or bunker gear, which includes a coat and pants, boots, helmet, hood, gloves, and self-contained breathing apparatus

Opposite: Firefighters work through heavy smoke to ventilate the roof of this fully involved apartment while others inside get a line on the fire. Although this was a training exercise, the conditions were very real and every bit as dangerous as an actual response.

Suburban fire departments often have the nicest rigs. This Pierce Saber from Gresham, Oregon, is no exception. From the beautiful graphics to the elaborate pump panel with a foam system and the hydraulic ladder rack, this engine did not come cheap. This upscale community, which overlooks Mt. Hood, exhibits pride in its fire equipment.

(SCBA). Together, this gear represents the personal protective equipment (PPE) to encapsulate the firefighter and provide protection against heat, smoke, fire, and noxious gases.

Each company has different responsibilities at a fire scene. The engine company must establish a continuous water supply and put water on the fire. Assignments vary between the first and subsequent engine companies arriving.

The truck company performs forcible entry, search and rescue, and ventilation. Ventilation involves opening the structure either vertically by cutting holes in the roof or horizontally by taking out windows and doors to allow the hot gases and smoke to escape from the enclosed interior areas. This reduces heat and acts to improve visibility for the interior crews. If the gases are not vented and the heat is not provided a means to escape, then companies are in danger of a flashover condition where the superheated gases build up and ignite. Truck personnel

may carry a set of irons, which includes an axe and Halligan tool, to forcibly open doors and windows. The Halligan tool or bar is a steel tool with a claw at one end plus a pick and wedge at the other end. It is used for puncturing, prying, and smashing, and it is especially well suited for forcing open doors. Other truck tools include pike poles for pulling apart walls and ceilings to check for hidden fire extension.

The rescue company can supplement the duties of either company by providing additional manpower or the services related to their specialized tools and equipment. The ambulance company may support the functions of other companies, they may be assigned to a rehab sector to monitor the physical condition of the other firefighters, or they may be responsible for rendering aid to injured civilians and firefighters.

Opposite: Firefighters are preparing to make an exterior attack on this apartment building to knock down the heavy fire on the balcony. One firefighter in full PPE is strapped in at the tip of the elevated ladder to direct the master stream. Some fire departments do not require the engineer/pump operator to wear full PPE, since that individual will remain with the rig and not go into the structure with the rest of the company. The rig is a 75-foot E-ONE Quint.

The Crescent Springs–Villa Hills Fire Department in Northern Kentucky is the proud owner of this 2007 American LaFrance custom aerial. The truck has an Eagle cab and chassis with a 110-foot LTI C34R-110 ladder. Truck 510 has a 2,000-gallon-per-minute (gpm) pump and carries 450 gallons of water along with 50 gallons of foam. The fire department had the rear body portion wrapped in a patriotic American flag decal on both sides, creating an eye-catching design. There is a prepiped waterway to the tip of the ladder to simplify deploying an elevated master stream at large fires.

Cicero Illinois Truck Company 2, due on a call for mutual aid, picks up at the scene of an extra-alarm fire in neighboring Oak Park. The Seagrave TDA was utilized as an elevated master stream. As the fire conditions diminish, the incident commander (IC) starts releasing out-of-town companies so that they can get back to service in their own districts. The driver and tillerman work in unison to steer the vehicle, since the rear wheels are controlled from the tiller cab.

BACKGROUND

The suburban areas surrounding major cities were originally residential communities or bedroom communities to house city workers. These towns often had some retail occupancies. In many cases, towns grew to incorporate a number of commercial businesses and, in some cases, they were developed to include an industrial base as well. In most instances, these areas were originally serviced by volunteer fire departments. As populations grew and the towns increased in size, it was not uncommon to see paid on-call firefighters and the introduction of career firefighters. Eventually, the economics of the growing suburbs forced many to build full-time career fire departments similar to those of larger cities.

Today, the suburban growth in many areas has pushed development even farther from the center of the cities. This has created a new outer ring of towns that are experiencing the same growing pains and

concerns that the inner ring of suburbs went through years ago.

The tax base for many suburban areas determines the level of service that a fire department can provide to the citizens. The disparity found from town to town can be minimal or enormous. This is reflected in the number and design of fire stations, the staffing and salaries, the equipment purchased, and the types of rigs. Often, suburbs have rigs that are bigger, fancier, and have more options than their larger city counterparts. This should come as no surprise, since the cities have larger fleets to maintain and therefore will generally be able to spend less for each so that their budget dollars go further.

All suburban departments, of course, are not staffed by full-time career firefighters. Some areas have volunteer departments, while others have some combination of the two. As demographics shift and *continued on page 20*

Firefighters who are also paramedics attend to the driver of a vehicle involved in a multiple-vehicle accident. The hydraulic tools in the foreground opened the driver's door, which was damaged. Firefighters carried a portable generator close to the car to provide power for the tools they were going to need. A backboard is ready for the patient, with a neck stabilization device already in place. Both fluids on the pavement are from the engine compartment.

Fire takes hold of the second floor of this suburban house after communicating from the garage. First-floor windows were taken out earlier to vent that floor, but concerns about the structural integrity of the roof caused the IC to order all companies off of the roof before they were able to initiate vertical ventilation. Without opening the roof over the fire while it was in the earlier stages, which would have given it a path to follow, the fire traveled where it had room to move before burning through the roof. This led it through the attic from the second floor.

RAPID INTERVENTION TEAMS

The RIT assembles with full PPE and an assortment of tools to execute a rescue at a position near the command post, or the front of the structure. They need to be observant of the ongoing operations, making note of the fire behavior and several means of egress, and paying attention to radio communications and the locations of companies operating at the scene. Here a five-man RIT remains vigilant near the main entrance to the fire.

Firefighters respond to fires with the responsibility of saving lives and property. If conditions permit, they will undergo an aggressive and thorough search to rescue trapped victims, along with mounting an aggressive attack to extinguish a fire. Most of the time, everyone gets out safely. There is a mantra in the fire service that states, "Everyone goes home," meaning all firefighters strive to return safely from each incident. But everything doesn't always go as planned. Firefighting is inherently a dangerous job. What happens if the firefighters become victims themselves? Who rescues the rescuers if they become trapped, lost, disoriented, or incapacitated inside a burning or collapsed structure?

In 2006, there were 106 line-of-duty deaths (LODD) for firefighters, and the number increased to 115 in 2007. Of the 2006 LODD, the greatest number came from stress-related injuries, followed by vehicle collisions, trapped firefighters, collapses, striking objects, lost or disoriented firefighters, and other

causes. Of the trapped and lost or disoriented deaths, more died in residential structures than in any other type of fire, and these typically are the most common type of occupancy that firefighters encounter.

The fire service needed to address the growing number of firefighters who were becoming lost, disoriented, or trapped in structure fires. The result was the creation of the rapid intervention team (RIT). Depending on department protocols, this may also be referred to as the firefighter assist and search team (FAST) or the rapid intervention crew (RIC). Regardless of the name, the concept was first introduced by the National Fire Protection Association (NFPA) in the late 1980s with the release of NFPA 1501. This provided national recommendations for a minimum of two firefighters whose sole mission at the scene of a fire is to maintain a posture of readiness in the event that one or more firefighters become lost, trapped, or disoriented, or are otherwise in distress.

There are also standards that require firefighters to stand by with full personal protective equipment (PPE) and a minimum compliment of tools when two or more firefighters go into a situation that places them in harm's way. This is referred to by some as "two in and two out," which is a federal mandate that was put in place by the Occupational Safety and Health Administration (OSHA) but is not meant to be the same concept as the RIT. OSHA 29 CFR 1910.134, which is part of the Respiratory Protection Standards, does not specify the depth and detail that is inherent in the necessary tactics and strategies to operate a RIT during a mayday incident, where a firefighter finds him or herself trapped, disoriented, injured, or running out of air.

A RIT company more commonly consists of at least four firefighters and must be ready to initiate search and rescue operations for firefighters transmitting a "mayday" call for help. If manpower is sufficient, a chief officer will also be assigned to the RIT.

The basic complement of tools for a RIT includes their PPE and self-contained breathing apparatus (SCBA), portable radios, hand lights, hand tools, search rope, stokes basket (a wire or plastic basket to carry a victim), supplemental air bottles and a mask to provide a downed firefighter

with additional air, saws, and a thermal imaging camera (TIC) that helps firefighters detect heat sources. Team members stage or stand-by close to the command post or the fire building so that they can deploy swiftly. Their main task is to anticipate a potential firefighter mayday by preplanning multiple-rescue action plans. They should inspect the building perimeter to find multiple entrances and exits. They may take on the responsibility to throw some ladders as an additional means of egress for firefighters if this has not already been done. They must monitor the radio communications so they know what is taking place and where companies are operating inside the building. In the event that a mayday is communicated from a firefighter, the RIT must have an initial idea of the building's layout, including entrances and exits.

In order to have an effective RIT, all firefighters must undergo specialized training for these duties, which include rescue techniques, recognition of

RIT training is essential to enable firefighters to be ready to handle the obligations that are part of accepting a RIT assignment. This involves, among other tasks, learning to work as a cohesive team to execute an effective rescue. One critical aspect is to train with real firefighters as victims instead of mannequins. This way, firefighters realize the nuances of real body mass and PPE.

fire behavior, breaching, and working as a cohesive team to execute an effective rescue. Although RIT training cannot possibly prepare firefighters to handle every unknown contingency that might occur at a fire scene, it includes procedures, skills, tactics, and maneuvers gleaned from actual incidents to allow firefighters the ability to learn from the events that others encounter. The training stresses that any one variable that changes in the rescue process can turn a 3-minute rescue into a 15-minute procedure or one with an unsuccessful outcome.

The guidelines for a RIT also state that they are not to be drawn upon to perform non–mission-specific tasks. Once they are deployed, additional resources will be assigned to assist them without compromising the fire-suppression activities that need to continue.

An effective RIT has all the tools at its disposal to effect an immediate rescue operation in the event of a "mayday." One important tool is a preplan of the location, showing the floor plan, building hazards, entrances, and exits. Here the RIT is assembled with its tools laid out on a tarp close to the building. All RIT members are in full PPE and are becoming familiar with the building's preplan, which was created for firefighters well in advance of this fire.

Someone torched this condominium building that was under construction. There was no life safety concern within the structure, which did not require much time before collapse. The Wheeling, Illinois, fire department went defensive with ground-level and rig-mounted master stream devices. There's not much to do at a fire like this, other than pour a lot of water on it until it goes out. The engine in the photo is a custom rig built by Sutphen.

continued from page 16
the inner-ring suburbs age, homeowners often move to the newer towns that have more to offer. This migration also spawns retail, commercial, and industrial development because many companies move where they can purchase land, expand into newer buildings, and be closer to workers. Subsequently, the tax base moves, and the older towns find it harder to collect sufficient tax dollars to sustain all of their municipal services. They begin making cuts across the board so they can pay their bills.

This phenomenon is common in many older areas of the Northeast and Midwest regions. As expanding areas build new fire stations, hire more firefighters, and buy new apparatus, older towns are closing firehouses, laying off firefighters, and subsequently finding it difficult to have sufficient manpower to safely and effectively fight fires and respond to other emergencies. It stands to reason that the older towns will have more fires, since they have older and more vacant properties poorer infrastructure, structures that are less fire resistant, and perhaps a fair amount of intentionally set fires. The newer areas will have more buildings with active and passive fire-control measures, which translates into fewer fires.

The Southwest is a prime example of an area that has witnessed explosive growth in the past decade. Many Southwestern suburbs have built large fire

Above: Two firefighters make preparations to put a ground-level deluge gun to work through the garage door at a fire involving an auto repair shop. The engine operator drags a 5-inch supply line for the gun. Overhead doors need to be cut open for access, since most are electrically controlled. The firefighters are in full PPE. The blue 2½-inch line is being deployed by a firefighter who is not visible.

When fire is in a garage, automatic garage doors impede firefighters' access to the fire. They have to cut the door to get water on the fire. Firefighters work together to gain entry whether they have to address the door or other obstacles. Here a saw is used to remove an obstacle.

Below: Firefighters in the basket of a Pierce tower ladder prepare to hit this apartment fire again with their deck gun. The fire started on the floor below and communicated from the balcony to the top floor. Conditions inside the building were untenable for firefighters at this point in the incident.

Above: The prepiped waterway on this Smeal aerial delivers an impressive volume of water, especially when the tip is so close to the building. After the main body of fire has been knocked down, this firefighter is directing the stream into a window to hit fire traveling along the ceiling. The metal box mounted to the side of the ladder houses a speaker, which allows a firefighter at the base of the ladder to communicate with the firefighter at the tip. Both positions have the ability to control the monitor. Any slight adjustment in the direction of the monitor that sends the water into the exterior walls results in an enormous splash-back and runoff.

Opposite: A firefighter opens the intake valve to let water into the pump from the supply line that was just charged by an engine. He has at least one hand line working off his rig, in addition to the elevated master stream on the aerial ladder not visible here.

departments with beautiful fire stations, large staffs, and top-of-the-line apparatus. They are progressive and have the ability to learn from mature areas. Some have the capability of beginning with a virtual blank canvas and can incorporate measures into the infrastructure in a much more cost-effective way than an older city. Traffic-control devices triggered by responding emergency vehicles to coordinate the signal to green for the rigs and red for others is but one example of a modern enhancement that can be included with initial construction. Sufficient water mains for fire hydrants, consistent fire suppression systems, and building access for the fire department are other issues for consideration when an area is starting from scratch. The ability to lay out optimal locations for fire stations throughout the town while residential subdivisions and industrial parks are being designed is also a great advantage. All of this combines to allow strategic planning to improve the fire department's ability to provide the highest level of service and protection to the residents.

Automatic aid agreements, which are relatively new to the fire service, expand the capabilities of any fire department by giving one department access to the resources of a neighboring department. In many areas, the fire company that answers an alarm may *continued on page 27*

continued on page 27

IS NEW CONSTRUCTION KILLING FIREFIGHTERS?

The need to have a thorough knowledge of building-construction methods and materials has always been an important aspect of a firefighter's training. This information, coupled with an understanding of the properties and behavior of fire, can mean the difference between a safe and successful outcome and a tragedy with loss of life, property, or both. There are inherent dangers that come with the job—some unforeseen and unavoidable, and others that can be minimized or controlled.

Efforts have been made to minimize the loss of life in the past several decades. Building codes have been bolstered, and fire awareness has increased with regard to educating the general public on prevention strategies. Firefighters have better PPE, tools, trucks, and training, and yet the advances have been tempered by some setbacks. Far too many civilians and firefighters still die annually as a result of fires.

The reasons for this loss of life are multiple and complex. Carelessness, inoperable smoke detectors, arson, and changes in some building standards are a few contributing factors. This section will explore codes and building construction, focusing on the concern today in the fire service about the detriments to firefighters with regard to lightweight construction.

First of all, there is good news. Building codes often require active measures to reduce, control, or extinguish fires. Sprinklers, smoke detectors, self-closing doors, and alarm systems have been incorporated into new construction of all types.

Everything today seems to come with tradeoffs, however, and building codes are no exception. For example, the addition or inclusion of sprinklers may appear to some to be a sufficient method to reduce the size of fires and minimize property damage and, therefore, the threat to lives. As it turns out, the added security gained by sprinklers and the additional expense of their installation can have a negative effect in that certain passive fire-protection or life-safety measures might be reduced or eliminated altogether. For example, the installation of sprinklers allows for a reduction in the hourly rating of some building components, which is the standard rating of a material to withstand fire as determined by the American Society for Testing and Materials. In addition, longer spans between fire walls, and even the elimination of selected means of egress, such as external fire escapes on high-rise buildings, are often allowed when sprinklers are incorporated into a building's design. Sprinklers and smoke-ejection systems are mechanical in nature and, as such, are subject to breakdown, maintenance issues, and human interface. Malicious or inadvertent actions could disable them. If the sprinklers malfunction, the remaining passive protective measures, which inherently make a building safer, may not be able to contain a fire.

In a perfect world, the sprinklers would contain and extinguish a fire, but building materials today contain large amounts of hydrocarbons, which burn at high temperatures. This intense heat may overwhelm the sprinklers' ability to extinguish a fire.

Although firefighters can never become relaxed or complacent when approaching a fire, they can have some confidence in knowing that certain older buildings were built to a level that will protect them from a premature collapse. Older buildings were heavier and often over-engineered. Designers and builders did not have access to the technical information and modern computer analysis/modeling. Today, engineers and designers have a greater understanding of the material properties, which allows them to know just what is needed to do a job. This increased knowledge translates into quicker, lighter, and less-costly construction. Some newer buildings, therefore, have less inherent fire protection for no reason other than they have less mass. This comes from a tradeoff between more efficient, lighter, and simpler construction as a result of the modern technical analysis of the materials versus the traditional methods, which used more materials, resulting in additional mass that, inadvertently, usually provided more longevity against fire and heat damage.

The construction trade is not to blame for all of this. The public prefers large, open spaces rather than the more compartmentalized designs of the past. The result is that walls and partitions that were never meant to act as supporting members previously could act as redundancies in the event of failures; now, spans are longer with little to fall back upon during a fire.

Sheathing materials have also changed. Some of this has to do with working toward a more environmentally friendly setting. Framing members for walls, floors, and roof decks used to be tied together with plywood, which made the stick framing more substantive and acted to distribute the load over a larger area. Now, more often that not, much of this plywood is being replaced with insulation products or oriented strand board (OSB), an engineered wood product. In these cases, plywood may be used only at corners and other locations where structural rigidity is required. All of this translates into the probability that any failure or weakening would potentially transfer the load to an area that is not strong enough to provide the needed support. Since the redundancies are not what they used to be, a collapse is possible in the event of a fire or other weakening event. The implications are that firefighters would have less time to make an aggressive interior attack to save a victim or protect the property and minimize the fire damage.

Another factor that contributes to the integrity of building construction is the use of unskilled and unqualified construction labor. As builders use more engineered components, the integrity and accuracy of their installation becomes extremely important. Again, without the redundancies of bulk in construction, each engineered product has specific requirements in terms of the appropriate types, sizes, and quantities of fasteners to maintain the strength, reliability, and integrity of the components so they perform as designed. Proper installation of these components is crucial to the building's integrity.

Still another tradeoff is the use of cold-formed, light-gauge steel rather than dimensional lumber in framing and metal joists. This non-combustible material has more than adequate strength and is well-suited for this purpose, unless it is subjected to loads in directions other than what it was originally designed to withstand, or if its cross-bracing and/or sheathing is compromised. Due to the inherent dimensional instability of the components, the stress of an atypical load can often translate into a quicker failure rate under these adverse conditions. A severe storm, for example, could apply forces in directions that the material might not handle. Traditional lumber, under these circumstances, may be able to maintain rigidity for a longer period of time.

More than ever before, firefighters have to be conscious of the type of building they are entering, and command personnel need to be ever more vigilant in monitoring the conditions they encounter and the safety of their personnel. As such, tactics have to be adjusted based on the knowledge of or suspicions about structural integrity when fires occur.

Fire departments should, therefore, utilize all opportunities to preplan buildings in their district and to inspect or view new construction so they can familiarize themselves for potential future events.

When the need arises, firefighters look for assistance wherever it is available. Here, bystanders were enlisted to help firefighters with the arduous task of laying several hundred feet of large-diameter supply line to fight a lumberyard fire.

Eight firefighters work together in the rain on a slippery, muddy slope to extricate the driver of this taxi that skidded into the ditch. A tow truck is also on scene and has attached a chain to the car to prevent it from sliding farther. Patients are placed on a backboard so they can be immobilized to prevent further injuries or discomfort.

A training instructor stokes a fire made up of wood pallets and bales of hay inside a building prior to letting companies inside. Firefighters will likely have to navigate their way through a smoke-filled building until they find the fire, perform a search, and hit the fire, all under the supervision and watchful eyes of a trained instructor.

Right: Firefighters from Woodbridge, New Jersey, encountered heavy smoke and fire conditions on arrival at a commercial structure. They immediately requested mutual aid, and the incident went to a third alarm as the fire rapidly advanced through the cockloft and vented through the roof. Initially, firefighters attempted to make entry through the rear before the IC directed operations to go defensive. *Michael J. Coppola, PublicSafetyPictures.com*

Below: A firefighter feels his way through heavy smoke to ventilate a window from the outside and check the immediate area inside. Many years ago, this same function would have been performed without the protection of SCBA, a Nomex hood, and without much more than a rubber coat and helmet.

continued from page 22
not belong to the city or town where the alarm originated, but rather to a neighboring fire department that was closer when the call was received.

MATURE INNER-RING TOWNS
Consolidations
The cities, towns, and villages that make up the inner ring of expansion around the major cities and metropolitan areas encompass a vast sampling of fire departments. Most of these areas are landlocked, which means that they have reached their maximum geographical size. This may not be the case, though, for the fire departments. Consolidation is an option for many areas where the public funding is no longer sufficient to maintain individual departments with either duplicate or insufficient services for each

The Glenwood Fire Department in the far south Chicago suburbs purchased two units in 2004 from HME. The engine and ladder work together at this storage barn fire. The intense heat requires firefighters to keep some distance from the building, which is a total loss. The blue lights along the length of the ladder are a relatively new design to enhance firefighter safety while climbing the ladder during nighttime operations and fire conditions with heavy smoke.

community. Merging two or more fire departments into one new fire department can be an efficient means of combining personnel, equipment, and other resources to provide a higher level of service to all of the areas originally covered by the separate departments. This can serve to increase staffing per fire company by reducing the total number of units.

A prime example of this type of organization was formed in 1999 when the towns of North Bergen, Weehawken, West New York, and Union City, New Jersey, collectively decided to merge their individual fire departments into a single department, which became the North Hudson Regional Fire & Rescue, the third-largest fire department in New Jersey. This new, larger department was able to implement a first-alarm response of four engines, two ladders, one rescue company, a deputy chief, a safety officer, and a mask service unit with supplemental SCBA for a report of fire or smoke in a building. If upon arrival it is determined that they have a working fire, they get an additional engine and a rapid intervention crew (RIC). A

Right above: When firefighters can no longer safely mount an interior offensive attack because of the threat of collapse, all activities turn defensive and are deployed from the exterior. Three elevated master streams are working at this vintage and extremely sturdy three-story building. Each ladder has a firefighter poised to direct the water for maximum effectiveness. Though all three aerials have prepiped waterways, only the small ladder on the far right of the frame is piped to the tip. The other two ladders are piped to the end of the third section of four-section designs. It is common to have the waterway stop short of the last, or fly, section so that it is not damaged if the fire department has to lay the ladder against the roof or parapet wall of a building.

Right: A firefighter uses a pike pole from a ladder to take out windows on the second floor of this single-family house. It is vital to the safety of interior companies that a building be ventilated to reduce the heat buildup, smoke, and hot gases to avert a flashover so crews can advance on the fire.

Firefighters from several towns were requested to provide mutual aid to the Glenview Fire Department in Illinois to fight this fire in a restaurant. Efforts to ventilate the building were complicated by the clay roofing tiles. As several firefighters maneuver a 2½-inch line from a ground ladder in the rear, others are positioning the aerial ladder to the roof. The firefighters on the turntable waiting to ascend the ladder are in full PPE, while the operator that will remain with the truck does not need SCBA.

response of this size would have been difficult with four separate departments. The merger is not without difficulties, however, both politically and operationally. This process is lengthy and involves labor agreements with the various firefighter unions, administrative considerations concerning chief officers, station closings, and the reassignment or relocation of companies. In the long run, consolidations like this one should prove to be more efficient, cost effective, and safer for firefighters due to an increase in response assignments. They should also provide improved service for the residents and property owners.

Upscale Bedroom Community

Fire departments in well-to-do bedroom communities find themselves in a unique situation. Often well paid, these firefighters generally see very little fire duty but find themselves providing a high level of customer service to the residents. Generally with less staff, these departments rely heavily on mutual and automatic aid from neighboring communities. Since they have fewer firefighters to respond, they need to beef up their automatic aid requests to get more people on the road in the event that they need the personnel at the scene. Many subscribe to the belief that it is easier to send resources home if they are not needed than it is to summon the extra help once they get on-scene. The realities of budget limitations require that staffing is adequate for 90 percent of the day-to-day operations. These firefighters often provide salvage and overhaul for residents that outstrips what a larger or busier department can implement. Devoting person-

Above: Firefighters in the foreground assemble their tools and equipment to assume RIT duties at this extra-alarm fire in Mundelein, Illinois, as a firefighter climbs an aerial ladder in the distance. Petroleum products burn hot and produce a thick black smoke. These factors combined to make this fire appear much worse that it was. Roofers and plumbers are notorious for starting fires due to the nature of the work they perform. This was the case here as roofers were working on the highest portion of this industrial complex on a hot summer day when the tar products ignited, creating a very dramatic fire. Once companies were able to get good water onto the roof area that was burning, the fire was extinguished quickly with minor damage to the building.

Left: It may seem odd to see a ladder going down inside this townhouse, but it accentuates the dangers associated with basement fires. By their very nature, basement fires are difficult and extremely dangerous to fight. Basements are confined spaces with generally only one means of egress and few, if any, windows. The lack of windows complicates ventilation, which means that the environment can be extremely hot with no place for the smoke to go, resulting in zero visibility. In addition, the free-burning basement fire is contained by the foundation and has nowhere to go but up. This presents serious safety concerns for companies entering the house not knowing the integrity of the floor. The image here illustrates a floor that burned through right in front of the rear sliding door. If firefighters entering through this door did not exercise caution and investigate the integrity of the floor, they could have fallen through the weakened structure into the raging fire below.

Fires in areas lacking hydrants are tough, but a large trailer park with blocks and blocks of trailers placed close together can be deadly. This is complicated further when the trailer park is nestled in the middle of an urban area that does not run water tenders. Fortunately, this fire occurred in the daytime, or it could have been considerably worse and caused multiple fatalities. The first alarm here sends more equipment than a similar response to an area with fire hydrants. After the first-in rigs exhaust their on-board water, they are reliant on backup companies either to lay long supply lines to a hydrant or, as seen in this photo, to nurse the initial rigs. Engine 11, which has a rear-mounted pump, is being nursed by Engine 9 and by the engine in the far distance. To conserve water, Engine 11 is mixing it with foam.

nel at a fire to move furniture and belongings together in a room so that they can be covered with tarps for protection from water damage is one example of a service offered by a small neighborhood fire department, as well as clearing out excess debris. They may also respond to help homeowners install smoke detectors or may make themselves available to provide blood pressure readings. These services may be taken for granted in an upscale community where fire duty is relatively non-existent and the residents expect more from the fire department.

The response to a reported fire might get two engines, one truck, one or two ambulances staffed with two firefighters/paramedics each, and a chief officer. Companies will generally be smaller if the departments are paid. Crews of three are most common. This would put anywhere from 12 to 15 firefighters on the first alarm. This restricts an initial attack or search to a crew of two, since the driver/operator will need to monitor the pump and secure a water supply. The second company, which could be the two-person ambulance, can continued on page 36

The sole occupant of this plane died after missing the runway at the Lee Airport and crashing in a field in Anne Arundel County, Maryland. The driver of Engine 2 connects a supply line from other units in the event that a fire breaks out. Due to the makeup of local fire districts, the city of Annapolis responded, as well as units from several volunteer companies in Anne Arundel County.

Above: From a safe distance, this firefighter directs the engine's fixed master stream on the seat of a fire. Standing on top of the rig in the winter with a strong wind blowing does not offer any protection from the cold.

Left: Arson was suspected to be the cause of this fire, which engulfed city hall within a block of the fire station. When firefighters pulled up, the fire had already made its way into the common cockloft and had free run of the nonsprinklered building. Two elevated master streams can be seen applying water, as well as the 2½-inch line from the street level. In the distance, a firefighter is attempting to vent the roof as heavy black smoke pushes from the eaves.

Right: A dramatic four-alarm fire in Oak Park, Illinois, engulfs the city garage that was loaded with vehicles and fuel. Companies went defensive soon after arrival due to the hazards within the building. A firefighter is settled in for the long haul manning the deck gun on Engine 603, which was made by Central States Fire Apparatus on an HME 1871-series chassis. The two small-brushed, stainless-steel doors at the base of the cab, as well as the matching door above the wheel well, are tubular compartments for the storage of extra breathing-air cylinders. *Steve Redick*

Opposite top: Though this pump operator seems very short, in reality he is standing in six inches of water that has run off from the firefighting operations. Squadzilla is a 1996 Pierce Lance pumper-squad with 750 gallons of water, on-board foam, a telescoping light tower, an air cascade system, and a full complement of rescue tools. The operator has just grabbed the remote control for the telescoping light tower and will light up the scene with 6,000 watts of light.

Opposite bottom: A nighttime fire in a lumberyard is a recipe for a huge loss to the premises. After the fire has been knocked down, there is a tremendous amount of overhaul to pull apart stacks of lumber to saturate the smoldering wood. While a firefighter visible in the distance works to pull the wood apart, the firefighters at the forefront maneuver a 2½-inch line to apply big water on the pile. Generally, only after the safety officer announces that the air is safe to breathe do firefighters use their discretion about wearing SCBA.

Left: Although the structure fire has been extinguished, this gas-fed fire cannot be stopped, since the gas shut-off at the meter has been compromised. Generally, the gas company can shut the gas off at the street connection to the house, but they were not able to get a crew to the scene to do so at this incident. The fire department opted to try a more difficult maneuver, where they dug through the debris to stop the gas at a point below the meter. Prior to putting these two firefighters to work, companies deployed three backup lines and two dry-chemical fire extinguishers, as well as an overhead mist from an aerial ladder. The intense heat from the fire damaged the neighbor's house by melting the vinyl siding and insulation.

This freak accident was complicated by a cargo of oxygen tanks that shot out as projectiles in all directions and reduced the van to rubble. Firefighters make sure that everything has cooled down and the fires will not reignite as other units continue to stand by. The driver, who is lucky to be alive, was pulled to safety before the explosions by a passing city worker.

The second floor of this house was fully involved when the first companies pulled up. They immediately called for mutual aid from several surrounding towns and were able to stop this fire before it burned the roof of the house. After the fire has been knocked down, the truck companies go to work performing extensive overhaul to prevent any hidden fire from spreading or reigniting. Firefighters apply foam to smother the hot debris. The firefighter is using a pike pole to pull down the soffits that are still smoldering. The only roof penetration is from the hole that was cut to vent the hot gases and smoke.

continued from page 32

assist the first company, put a second line to work, or function in an EMS posture if there are injuries. The truck company would most likely begin ventilation, perform a search, or be responsible for forcible entry depending on the circumstances. Some suburban areas have the luxury of staffing a truck with four, since the truck covers a greater area than any engine company and becomes a source for additional personnel; conversely, it is for that same reason that some areas staff a truck with only two, since fires are down and it is a supplemental piece of apparatus. These departments feel that if the aerial is actually needed at a working fire, they will bolster the company with personnel from other rigs as long as they have someone to get it to the scene. If there is a fire of any magnitude, it will not take long for the incident commander (IC) to request additional units immediately.

GROWING OUTER-RING COUNTIES

As the inner ring of communities surrounding Chicago grew and matured, the communities that ringed these towns started to expand. The farms and vast open lands began to give way to housing developments, strip malls, shopping centers, and business centers. At the same time, the communities had volunteer fire departments or fire districts that levied their own taxes. These departments saw their call volumes increase and their abilities to get sufficient
continued on page 40

Two firefighters handle an attack line in Lincolnshire, Illinois, at this fully involved attached-garage fire. There is still quite a bit of fire to extinguish now that the roof and walls have begun to collapse.

Above: Big fires need big water, and big water often means many lines. In this image, three engines can be seen with lines off, and a fourth is directly off to the left of the frame—and this represents only a fraction of the scene. Every length of hose is labeled with the name of the fire department and the rig where it belongs. As the IC determines that individual lines can be shut down, the long process of picking up begins. When a charged line rests on top of a line being put away, firefighters have to lift the heavy one first before they can proceed. All sections of hose get drained and then rolled up or dragged back to the rig so that the hose bed can be repacked to get the rig back in service for the next fire.

Right: A firefighter pulls a preconnected attack line off the rear of a rig. He has hold of a strap that is around a bundle of hose to simplify getting it out of the hose bed for rapid deployment.

THERMAL IMAGING

Firefighting and the fire service in general remain steeped in tradition. Whether this refers to the color of trucks, the generations of family members in the fire service, or that the new guy is always the last in line for food, there are many types of traditions. The fire service can also be very resistant to change. Improved PPE, new tools and equipment, automatic fog nozzles, compressed-air foam, the incident command system, training, computer-aided dispatch, trunked radio systems, and bigger rigs are all concepts that have run into resistance during their introduction to the fire service. Each of these, as well as many other facets of firefighting, has added significant advancement in safety or fire suppression.

One example of an undisputed technological advancement was the introduction of thermal imaging to the fire service. This technology was initially available only to the military, and it was very costly. It was first introduced to the fire service in the 1980s. In the 1990s, design changes made units less bulky, production subsequently increased, and prices decreased. Originally, when the technology was very costly, a department might have just one unit that was carried to the scene of a fire in a chief's vehicle.

Today, thermal imagers (TI) or thermal imaging cameras (TIC) are commonplace in most fire departments. Many departments can

As firefighters man a big 2½-inch hand line outside this house fire, an officer surveys the house with a TIC to check for hot spots or hidden fire, since everyone was instructed to evacuate the structure. The TIC uses infrared technology to show heat on the monitor.

afford to have a TIC assigned to each frontline fire-suppression unit. Most TICs are handheld devices, although there are some that can be helmet mounted, allowing firefighters full use of their hands while retaining the thermal imaging capabilities. Early TICs were bulky and remained on the helmet at all times, but newer models are much more compact and can be interchanged between firefighters.

The TIC uses technology that allows firefighters to detect heat through smoke and walls, enabling them to locate unconscious victims or to find the seat of a fire and detect structural dangers. Hot spots behind walls and above ceilings can be detected, eliminating the need to punch holes in the walls and pull down ceilings in unaffected areas. The TIC may provide valuable information during size-up or initial evaluation of the incident, which can assist the IC in determining a strategy and developing an incident action plan. A TIC can save a great deal of time by helping to pinpoint a concentration of heat within a particular area of the building, especially in large commercial or multistory structures. A TIC also proves valuable during search and rescue operations by reducing the amount of time it may take to use otherwise standard search techniques.

This small, handheld TIC makes it both easy and convenient for the firefighter to make sure that all pockets of fire have been extinguished. Firefighters from the truck company already performed basic overhaul. Although it does not appear that there is much left of this room, it is still prudent to perform this inspection, since the structure is still standing.

The TIC has become an integral tool for firefighters during searches for victims or fire, and in the overhaul mode to check for fire extension. After the walls and ceilings have been torn down in the areas that were obviously affected by the fire, a firefighter following up with a TIC can confirm whether they got it all or if there is still more work to be done.

Winter fires in the Midwest pose challenges to firefighters, including deep snow, ice, and cold temperatures. The water freezes on all surfaces, making conditions slippery. Here, some firefighters can be seen inside the crew cab taking refuge from the cold. The firefighters on top of the engine are preparing to put the deck gun into operation, since the fire attack is in a defensive mode where all firefighting is done from the outside. The rig is a Pierce Saber pumper.

continued from page 36

personnel out the door decrease. Even a steady one call per day could put a strain on the volunteers, their jobs, and their families. Many firefighters had jobs out of town, which meant they were not available during the daytime hours. In the past, departments did not call for help unless they really needed it, since automatic aid is a relatively new concept. In addition, as much as EMS is an integral part of most fire departments today, medical transport was always provided by the funeral homes before the emergence of modern-day EMS.

Most of these departments began hiring paid staff members in the 1970s and 1980s. This was initially met with some animosity from volunteers, since the paid firefighters trained more and embraced newer tactics, tools, and procedures. Today, practically the entire region is made up of full-time career firefighters. They regularly utilize automatic aid agreements to ensure adequate resources will be on hand to guarantee the safety of all firefighters and provide the maximum service

available to residents. This is, for these departments, the best utilization of available resources.

BUDGETARY STRUGGLES

Fire departments in mature communities can face financial stress when voters limit tax increases. This can result in layoffs, station closings, and a reduction in safety for firefighters and residents alike. It is a problem that is not going away. Fire departments have to rethink the way they do business and provide service to the residents. When budget cuts are deep, there becomes a very real concern for the safety of firefighters. Without enough personnel on an alarm, their ability to put out a fire quickly and simultaneously search for victims becomes extremely difficult. However, if they push themselves to work heroically, as do so many dedicated firefighters across the country, it is possible that the citizens and civic leaders will not understand that the fire department needs more resources. Sadly, sometimes only a tragic event can initiate change.

Left: Fire vents from the second-floor windows and eaves of an apartment building. The outer windows have already broken, and the center pane is about to blow out. Smoke will find its way through any space that is not airtight.

Below: Firefighters in Colerain Township, Ohio, use their E-ONE Typhoon pumper at a fire in a small abandoned house. There were no life safety concerns here so they went defensive. They laid a 5-inch large-diameter supply line from the hydrant to feed the engine in the driveway.

CHAPTER TWO

URBAN FIREFIGHTING

Big-city municipal fire departments share as many similarities as they do differences in tactics, training, apparatus, staffing, philosophies, and attitudes. It would be difficult to compare cities the size of New York, Chicago, Los Angeles, Boston, Philadelphia, San Francisco, and Washington D.C. with smaller cities like Yonkers, Manchester, Newark, Buffalo, and San Jose on a basis of size, fire duty, number of calls, and manpower.

On the basis of sheer size alone in all categories, no organization compares to the Fire Department of New York (FDNY). If this department were divided into a separate department for each of the five boroughs, they would each be some of the largest fire departments in the country. With that said, the FDNY has 15,607 members. It runs 347 engines and trucks and had an annual call volume in 2006 of 246,721 non-EMS responses. By comparison, the next largest city fire departments are Chicago with 5,060 members, Houston with 3,835 members, Los Angeles City with 3,594 members, and Baltimore City with 1,679 members. In terms of engines and trucks in service, Chicago leads with 159 engines and trucks, Los Angeles City has 148, Houston has 125, and Baltimore City has 58. Run statistics are difficult to scrutinize because there is no consistency of reporting for all fire departments. One source listed total non-EMS responses in 2006 for these departments as 278,408 for Chicago, 335,278 for Houston, 126,942 for Baltimore City, and 63,198 for Los Angeles City. This small sampling shows that the ranking of number-two city in the fire service differs when taking into account the number of personnel, the amount of frontline apparatus, or the number of non-EMS responses.

In terms of the U.S. fire service, ranking the size of a fire department goes beyond the cities, since there are very large county-run fire departments that do not represent any one particular city. Large county fire departments are prevalent in many areas of the United States. The Los Angeles County Fire Department is the largest in this category with 4,656 members, 192 engines and trucks, and a reported 52,561 non-EMS calls in 2006, while the Miami-Dade County Fire and Rescue Department has 2,329 members, runs 56 engines and trucks, and had 52,833 runs. Other large county departments are in Anne Arundel County, Prince George's County, and Baltimore County in Maryland; Orange County, Florida; plus Orange County, Santa Barbara County, and San Bernardino County in California. The makeup of county departments varies greatly. Some provide coverage for unincorporated areas, while others encompass cities and towns that do not have their own municipal fire departments. Small towns or cities may opt to disband their own fire departments for budgetary reasons and contract for the services of a county department that can offer fire protection more cost effectively. If a municipal or volunteer fire department fails, is unable to pay their bills, or cannot continue to provide services to the residents, some laws require the county to absorb the resources and district into the area covered by a county department.

County fire departments can be responsible for vast amounts of land. As a result, they may have to protect very different areas, including urban, suburban, rural, and wildland regions. In addition, their districts may cover residential, commercial, and industrial areas. An example of this can be found in

Opposite: Big fires require big water. After the fire vented through the roof of this commercial building on Chicago's Northwest Side, a firefighter across the street put the deck gun to use immediately. The heat was so intense that the engine placement was almost too close for safety. Once the fire attack goes defensive, all firefighters stay outside the structure and put multiple master streams into operation until the fire is knocked down. Then they shut down the master streams and move in with hand lines to finish the job. Chicago Engine 126 has several lines off at this 3–11 alarm fire on the city's south side. Burning out of control is a supermarket. Farther down the street, one of several aerials can be seen through the smoke deploying an elevated master stream.

Above: Engine 65, located in Manhattan, sits at the end of the block at a hydrant waiting for instructions from the IC down the street. The company is at the building, but the chauffer remains with the rig. Life goes on in Manhattan as workers barely notice the commotion.

Right: The Baltimore City Fire Department is on the scene of a third-alarm fire in a row house on the east side of the city. The nature of these structures poses a big risk for extension to units on either side, but companies were able to contain the fire to the unit of origin. Truck 1 is positioned to put its main on the roof, if needed. Both trucks are parked at opposite corners of the building, allowing each to access the front and one side.

the Miami-Dade County Fire Rescue Department, which provides coverage for vastly urban areas but is also responsible for much of the less-populated area in the southern part of the county with large, open areas and no fire hydrants. One end of the county has urban staffing and tactics, elsewhere they deal with suburban homes and commercial districts, while the other end has less staffing and runs tankers or tenders that carry larger amounts of water to sustain firefighting operations without a fixed system of fire hydrants.

The Los Angeles County Fire Department has grown in the last 10 years by assuming fire protection for many municipal departments that have closed their own fire departments and through widespread, rapid growth of the undeveloped areas.

In Maryland and Virginia, most county departments include career firefighters and volunteers. As the volunteer organizations lose their ability to attract sufficient member numbers or provide adequate responses during weekdays, the void is filled by an expanding career force that continues to grow and support the increasing demand. Fairfax County, Virginia, is a prime example of this. Although more than a handful of the county fire stations are owned and lettered for volunteer companies, the rigs are staffed with career personnel during the days. Most of the volunteer companies raise their own money to buy the fire apparatus of their choosing, while the stations that are strictly career houses have equipment purchased through the county's normal procurement procedures. Apparatus purchased by the volunteers may not match the county apparatus in style or color, since the volunteers have their own colors and lettering to reflect their traditions and history.

To complicate the question of size just a bit more, the California Department of Forestry and Fire Protection (formerly known as CDF, now known as

Phoenix companies approach an early morning fire in an abandoned structure. Almost all of the frontline engines in Phoenix have the mid-engine design with a rear pump, as seen here. The department opted for this design, giving the crew more room in a much quieter cab, in addition to moving the pump operations farther from the noisy motor. The rigs also feature roll-up doors so that all of the compartments can be opened without creating any obstructions in the street.

Burlington, Vermont, purchased several engines and a mid-mount tower ladder from E-ONE in 2003. These units here are on-scene at an industrial complex. Firefighters used the tower to get on the roof for ventilation. The tower has a prepiped waterway to the bucket, which allows them to use a master stream or to attach hand lines to work on the roof. The engine pictured here is supplying the tower with a 5-inch supply line.

CAL FIRE), which is a state-run fire department, is responsible for upwards of 15,000 career, volunteer, seasonal, and inmate firefighters at peak times of the year and 1,133 engines and trucks spread throughout the state of California.

The ability to mount an aggressive attack on a fire depends on the resources available to the fire department. While the FDNY can put 60 firefighters on-scene in a matter of minutes, the Denver Fire Department will arrive with half that number, while Miami-Dade County Fire and Rescue will have even fewer on the first assignment. These dissimilarities in resources can greatly affect the outcome of an event in terms of lives saved, firefighter safety, and property loss. The jobs that need to be executed on the fire-ground do not change, it's just the when and how a department is able to perform the jobs based on the assets at their disposal.

SINGLE-FAMILY RESIDENTIAL FIRE
Miami-Dade Fire Rescue
When Miami-Dade Fire Rescue responds to a single-family residential fire, they send three suppression units and one medic rescue. They run engines with 65- or 75-foot aerial devices, quints, and three-person ambulance units, which carry extrication equipment in addition to the medical supplies. Staffing for a suppression rig is four. All personnel are cross-trained to perform the functions of an engine company, a truck company, or a rescue company. This will put 15 firefighters on the first assignment. The first company attacks the fire; the second company augments the water supply and either does truck work or puts a backup line in place. The third company could lay a supply line, go to the rear for a secondary egress, perform ventilation or forcible entry, or disconnect the utilities.

Left: This hose roller rests on the edge of a cornice or parapet wall to assist firefighters in pulling rope or hose up to the roof of a building. The hose roller protects the hose or rope from scraping along the edge of the building, which could cause damage to either. Here, a 2½-inch hose line is being hoisted up onto the roof from the street below. The line will be deployed from the rooftop of a building across the street or adjacent to the building that is on fire.

Below: The Miami-Dade Fire Rescue Department purchased 15 of these rigs from Rosenbauer. The rigs have Spartan Gladiator chassis with Evolution cabs and 60-foot RK aerials. They have a 1,500-gpm pump and carry 500 gallons of water and 30 gallons of foam. These are quints and were designed to replace the department's aging fleet of TeleSQURTS. Aerial 2 is the busiest company in Dade County.

Right: Harsh weather conditions cannot deter firefighters from performing their duties. Severe heat or extreme cold both present unique challenges. Everywhere in the country, firefighters have to deal with high heat or humidity during at least parts of the year, while those working in the northern and Midwestern areas also incur sub-zero temperatures and bone-chilling winds. In all cases, firefighters need to adapt to the current conditions. Here, a group of Chicago firefighters works to disconnect lines from a large manifold after an extra-alarm fire in frigid conditions. Since everything is frozen and covered with a thick layer of ice, they use a sledgehammer in combination with brute force to break the fittings free. Out of the frame, a firefighter stands ready with a torch.

Opposite: Detroit firefighters go to work at a second-alarm fire in an abandoned warehouse on the city's southwest side. Two firefighters in the bucket of Ladder 7, which is operating as a platform, are getting into position to put their dual elevated master streams into operation. The 95-foot Sutphen tower ladder that they are assigned has a unique design, with large monitors situated on each side of the bucket. In front of Ladder 7, Ladder 8 already has their master stream working. Ladder 8 is a 100-foot Pierce TDA. Detroit firefighters regularly see a large number of fires. A fire of this magnitude in most other cities would be anywhere from a fourth alarm to an eighth alarm. *Steve Redick*

Detroit

The Detroit Fire Department, like those in many older cities in the country, is experiencing budget cuts and monetary shortfalls. Due to manpower shortages, Detroit is required to temporarily close 10 fire companies per day around the city to conserve resources. Their response to a fire in a single-family dwelling is called a box alarm and brings three engines, one ladder, one squad, and a chief to the scene. They staff each suppression rig with 4, putting a total of 21 at the incident. Typically, upon arriving and confirming a fire, the first engine will announce, "We're stretching," which means they will drop a skid load (2½-inch supply hose) and a bundle (1¾-inch attack line) in front of the building and then lead out to a hydrant. The supply line will have a wye adapter, which allows the second engine company *continued on page 52*

continued on page 52

COMMUNICATIONS AND DISPATCH

Citizens rely on the fact that when they have an emergency, they can place a call for police, fire, and emergency medical services (EMS). Years ago, every emergency request required a different seven-digit phone number. If the number dialed was incorrect or reached the wrong department, then the call would require additional time to get to the proper agency that would respond to the call for help.

HISTORY

In 1967, various departments of the federal government agreed that there should be a single number nationwide for reporting emergency situations. The Federal Communications Commission brought the issue to American Telphone and Telegraph (AT&T), which, at the time, was the sole provider of phone service nationwide. AT&T proposed the three digits—9-1-1—as a national number for requesting emergency help. This number set was selected because it was easy to remember and dial and because it was a combination of numbers that was not in use anywhere in the phone service as an area code, service code, or prefix. Congress passed legislation in 1968 to ensure that only this series of numbers would be used for this purpose and for no other purpose. By the end of the twentieth century, roughly 93 percent of the United States had some type of 9-1-1 service.

As technology has continued to advance, the abilities of the 9-1-1 systems have been enhanced to provide information about the caller and the location, and any other information that can be uploaded into a database to assist emergency responders. Modern 9-1-1 systems can triangulate via cell towers to locate a cell phone caller, whereas earlier systems only had the tools to trace hard-wired phones. Systems can also be programmed so that the emergency services dispatch center can send voice or text messages to 9-1-1 subscribers en masse, enabling them to notify many people of impending danger.

In many cities and counties, 9-1-1 call centers and emergency radio dispatchers are located in large regional or centralized facilities in an effort to improve the level of service to the public and take advantage of cost efficiencies while reducing redundancies. These facilities may or may not combine police and fire together, and they might cover multiple municipalities or departments.

It would be impossible to cover all the different 9-1-1 procedures nationwide. Different staffing, different technology, and other concerns create a wide variety of response systems. One example of a state-of-the-art system exists in Chicago.

CHICAGO OEMC

The City of Chicago Office of Emergency Management and Communications (OEMC) was created in 1995 to coordinate emergency communications, which includes receiving 9-1-1 calls for help and dispatching police, fire, and EMS. The capabilities located within the state-of-the-art Chicago facility, though, encompass far more than these emergency services. The OEMC is a first-of-its-kind organization with a facility that combines surveillance throughout the city with 9-1-1 services. This is also the location of the Joint Operations Center, where multiple agencies and city departments would gather in the event of a large-scale event to provide a unified command post, allowing all participants to work together for monitoring or mitigating an incident.

Another aspect of this facility is the monitoring of remote cameras located throughout vital areas of the city. These cameras provide real-time coverage both of specific locations and broad areas. This includes traffic cameras at intersections and in school zones thoughout the city, police cameras in high-crime areas, security cameras in every public school, and a saturation of cameras in the downtown sections. Everything is linked by a network of copper and fiberoptic cables that is owned and maintained by the city of Chicago. This infrastructure has positioned the city well for future growth and expansion of monitoring and communications systems.

The U.S. Department of Transportation has selected the Chicago OEMC to pilot a program with the next generation of 9-1-1 technology, which will allow the center to receive photos and text messages from cellular and mobile devices, as well as being able to respond to the text messages. This new capability is meant to capitalize on the fact that so much of the population has these devices and can transmit information to help the authorities mitigate or find emergency events. Text messaging allows a hostage to communicate with the police without having to speak out loud. The cameras can help determine someone's location or provide vital intelligence to the police during the commission of a crime.

9-1-1

The Chicago 9-1-1 center answers between 17,000 and 20,000 calls for service per day, with almost half originating from cell phones. The majority are requests for the police department, and the space allocated for the police is representative of their share of the call volume. The fire and EMS call takers are responsible for all calls requesting medical assistance, reporting fires, and accidents, and other calls for aid. The call takers answer the phones and simulteanously transfer the information to a dispatcher, who immediately notifies the appropriate fire department companies to respond. Dispatchers are separated between fire suppression and EMS, with a supervisor as backup to oversee all operations. It is the responsibility of

the dispatchers to monitor radio traffic, fulfill the requests of units on the street, and verify that they receive all pertinent information that is transmitted via computer terminals in the fire houses and in each rig. Although the information that dispatchers need is provided by the computers, there is no replacement for the experience and knowledge possessed by seasoned veterans who maintain a thorough knowledge of the city and available resources to ensure that every situation is handled in the most efficient and professional manner.

Call takers and dispatchers can view, via the computer monitor, the caller's location and other information relevant to the incident. All vital information, which fire companies or fire prevention bureau inspectors find during field inspections, is uploaded into the database and will be available to firefighters in the event of an emergency. Specific hazards or the location of people with disabilities are important details that the fire-alarm office can relay to responding units via the computer messages or over the radio.

Fire-alarm operators can monitor the location of fire companies with global positioning data that is transmitted from each vehicle. This information allows for the human interface to help ensure the best use of resources when companies are on the street and not in quarters.

Emergency dispatch and call centers have come a long way since the introduction of 9-1-1 systems and the integration of modern technology into the job.

Within the Chicago Office of Emergency Management and Control is the 9-1-1 center that is shared by the fire and police departments. The Chicago Fire Department dispatches on the Main and Englewood frequencies, which divide the city's companies between north and south. Prior to the opening of this installation, the Main and Englewood Fire Alarm Offices were in separate locations.

Day and evening shifts require a minimum of three supervisors, seven dispatchers, and six call takers. The call takers work citywide, and the dispatchers are divided between Main and Englewood for fire and EMS. Three monitors in front of the dispatchers provide maps, unit status, and active events. The radio communications have touch-screen monitors, and the large-screen projection illustrates active events on a citywide basis at a glance. Each console has a tower with four lighted segments. One signifies that the operator is transmitting via the radio, and another indicates that the operator is actively on the phone. Lastly, the large map with lights on the wall offers an indication of the availability of all engines, trucks, battalion chiefs, and deputy district chiefs. Green signifies the engine company is available, and amber represents an available truck company.

On May 4, 1988, the Los Angeles City Fire Department responded to 9-1-1 calls reporting a fire at the First Interstate Bank Building, a response that was delayed by at least 15 minutes because building security reset numerous internal alarms and then sent maintenance to investigate before notifying the fire department. The fire started on the 12th floor of this 62-story building and eventually destroyed 5 floors. The fire was extinguished in just less than four hours. One building maintenance worker died in an elevator after being sent to investigate the fire. *Rick McClure, LAFD*

continued from page 48

to attach the attack line to this supply. The ladder company will park in front, sending firefighters to the roof, while others throw ground ladders. If the ladder arrives first, they will pull a line and go to work from their 250-gallon on-board water tank. Vacant dwellings might be hit quickly with a deck gun from the second engine pulling in behind the ladder. A fourth engine can be requested to assume

RIT duties. Engine drivers do not become part of the attack company; they remain outside.

If needed, a second alarm brings three additional engines, one ladder, a platform aerial, an additional squad, and two chief officers.

HIGH-RISE BUILDING FIRES

Fighting fires in high-rise buildings presents many challenges. Whether the building is 10 stories tall or 110, firefighters have to fight the fire from the inside if it is above or outside the reach of aerial ladders. This means that all of the equipment has to be carried into the building and transported to the upper floors before going to work. This is also complicated by the fact that the building occupancy can be enormous during the daytime in an office building and at night in a residential building. Firefighters need to be able to communicate with the tenants and determine if the best plan of action is to evacuate the building or have the occupants remain in their units, which is known as a "defend in place" strategy. Newer buildings have centralized public address systems that the fire department will use to broadcast instructions to the residents or tenants.

Resource requirements to fight a high-rise fire are huge. The balance between assigning resources to fight a fire in the incipient stages before it is able to grow and allocating the resources to evacuate civilians, not attacking the fire until everyone is out of harm's way, is delicate. These fires require huge numbers of personnel, lots of equipment, and expert management in order to achieve a successfully coordinated and controlled outcome. First-arriving units determine the attack stairwell that will be used by firefighters attacking the fire, as well as the evacuation stairwell, which will be used for the evacuation of tenants or residents. They also will take control of the elevator operations and the public address system.

Due to the size and complications of a high-rise fire, the Incident Commander (IC) has to predict the need for additional resources and request assistance early in the incident. Time is an important consideration at high-rise fires, since it will take longer to get water on the fire than would be the case in a conventional structure fire. Firefighters can potentially get water on a house fire within minutes by using a pre-connected hand line. At a high-rise fire it might take 10–15 minutes by the time companies reach the floor below the fire, connect to the standpipe (the internal water supply for the fires), and advance up the stairs to attack the fire.

Few fire departments have the resources needed for a prolonged event in a high-rise. In order to adequately

Left: Box 22-3348 was struck at 2475 Southern Boulevard in the Fordham section of the Bronx for a fire on the 12th floor of a residential high-rise building on August 21, 2006. The building had 31 stories on one end and a lower 14-story section where the fire occurred. This second-alarm fire was held to the apartment of origin, and there was one civilian injury. Residents on and above the fire floor were evacuated, while those on the lower floors were advised to stay in their apartments. *Matt Daly*

Above: The FDNY has both the ability and the luxury of being able to amass almost 60 firefighters to a reported fire in a high-rise. As is most often the case, many companies stage in front, awaiting an assignment. The vast majority of these incidents are minor incidents or false alarms. The stokes basket laden with tools represents the FAST.

and safely sustain firefighting operations and searches in an active high-rise incident, a fire department should have three to five times the resources for support functions. Major fires in high-rise buildings may require 100–200 firefighters in addition to a large command and support system.

Prior to the 9/11 terrorist attacks, several major high-rise fires served as wake-up calls to the fire service due to the magnitude or the lives lost. These resulted in tactics, training, and response changes for the fire departments affected, and for many who chose to incorporate these lessons into their own plans of action. Unfortunately, the hard lessons learned were not adapted across the fire service, and

some departments are ill-equipped to handle a similar emergency in their own cities.

HISTORICAL INCIDENTS

On May 4, 1988, the Los Angeles City Fire Department responded to a fire at the First Interstate Bank Building, which resulted in the destruction of floors 12–16. Numerous smoke alarms activated in the building, but security personnel reset them. Several minutes later, when additional smoke detectors were activated, a maintenance employee went by elevator to investigate the alarms on the 12th floor and subsequently died in that elevator from inhalation and thermal injuries. The fire department was

Right: Pictured here are the remains of the office suite on the 12th floor of the Cook County Administration Building at 69 West Washington Street in Chicago. The extent of the devastation is evident in the twisted and charred rubble that was a ceiling, office furniture, workspaces, and cubicle walls. It is hard to imagine entering this inferno to navigate around the furniture and other debris with zero visibility and extreme heat.

Below: This is another view of the office suite in the Cook County Administration Building, as seen from the main entrance. On the right, stanchions are visible in what was the public waiting area. In the foreground are the remains of filing cabinets and other furniture. In the background, the 12th floor windows have been boarded up.

first notified of the fire by calls to 9-1-1 from outside the building, seven minutes after the security personnel received the initial alarm. The fire department fought the fire for close to four hours, committing 64 fire companies and 383 firefighters.

On February 3, 1991, the Philadelphia Fire Department was dispatched to a fire on the 22nd floor of the One Meridian Plaza Building. The fire would eventually burn for 19 hours and consume 7 floors of this building, being stopped at the 30th floor by automatic sprinklers, which had not been installed on the lower floors. Three firefighters lost their lives at this fire, and 24 others suffered injuries. This was a 12-alarm fire that utilized over 300 firefighters at the scene with 77 fire suppression and specialty units.

On October 17, 2003, the Chicago Fire Department received calls for a fire on the 12th floor of the Cook County Administration Building. At one point, 13 civilians were trapped in the attack stairwell. Six of them were overcome by smoke and died. Chicago had 33 fire suppression companies and 23 ambulances assigned to this fire, representing roughly 185 firefighters, paramedics, and chief officers.

As a result of each of these fires, municipalities and fire departments throughout the country had the opportunity to learn from what happened and to take action to ensure that the mistakes or problems that hampered those fire departments would not be repeated. These problems were relevant at the One Meridian Plaza Building fire, including the need for greater control over evacuation and attack stairwells to safeguard civilians during evacuations. Another lesson was the need to use smooth bore tips instead of automatic nozzles with a standpipe to avoid nozzle-clogging debris and to combat the fact that standpipes are low-pressure systems. Many departments took advantage of these lessons, though others did not. Complacency, with the thinking that a similar incident is unlikely elsewhere, may be the strongest obstacle for many cities to overcome.

COMPARISON OF RESOURCES AND RESPONSES IN URBAN AREAS

FDNY

When the FDNY gets a reported fire in any high-rise building, they initially send a standard structural response consisting of three engines, two trucks, and a battalion chief. Upon confirmation of a working fire, the first arriving company will transmit a "signal 10-75," which brings another engine, an additional truck, a second battalion chief, a rescue, and a squad company. At this point, the FDNY has between 52 and 56 firefighters or chief officers at the incident. They staff rescues, squads, and trucks with a total of 6, and engines can have 5 or 6 on board. Each battalion chief in New York is assigned an aide. If the fire is significant, an "all hands" will be transmitted, which adds a division chief and a rehab unit to the scene.

FDNY supplements the standard structural response when there is a working fire in a residen-

Right: On February 26, 2006, FDNY firefighters responded to Box 33-3553 for a fire on the 24th floor of one of the Tracy Towers on West Mosholu Parkway in the South Bronx. The 41-story building is one of two twin structures, the tallest in the borough. When firefighters were in the involved apartment, the windows blew out, and 50-mile-per-hour winds created the equivalent of a blowtorch in their direction. Three firefighters suffered burns, and several others incurred minor injuries, as did two civilians before the third-alarm fire was brought under control in about an hour. The fire self-extended to the 25th and 26th floors directly above the unit of origin. *Matt Daly*

American LaFrance built eight custom Eagle pumpers for the San Francisco Fire Department in 2005, which were delivered for the fire department's centennial celebration in 2006. These were painted in a color intended to replicate the color used by American LaFrance in 1906. Subsequent and previous equipment purchased had a brighter red paint with more white on the cab. The centennial logo can be seen on the rear cab door. These engines carry 500 gallons of water, 30 gallons of foam, and a 1,500-gpm pump.

tial high-rise building. For this type of incident they would transmit a "10-77" at any point in the process outlined above. The total response for the "10-77" is five engine companies; five trucks; six battalion chiefs, including the duty safety and rescue operations battalion chiefs; one division chief; one rescue; one squad; and the field communications unit, which is a mobile communications center with an officer and two dispatchers.

Once on-scene, the first and second engines are responsible for the first attack line, originating from the standpipe one floor below the fire floor. The third and fourth engine companies team up to put the second hand line to work either as a backup to the initial attack crew or as a line on the floor above the fire. This line originates two floors below the fire floor. The fifth engine is assigned as a certified first responder unit to attend to injured firefighters or civilians inside the building. It is responsible for getting these patients to the EMS crews staged elsewhere.

The first truck is responsible for search and rescue of the immediate area surrounding the fire. The second truck will meet with the first truck and search the fire floor before searching for victims in the attack stairwell. Truck company operations are not carried out with the entire crew performing the same task. FDNY assigns specific tasks for each riding position on the rig. The inside team includes the officer, the "can" (the firefighter who carries the extinguisher), and the "irons" (the firefighter with the forcible entry tools). The outside team is made up of the chauffeur, the outside vent (OV) position, and the roof position. The OV and the roof firefighters assigned to these trucks will make their way to the roof to initiate ventilation in the attack stairway. The third truck handles the roof assignment if the first two trucks are unable to accomplish this; otherwise, the third and fourth truck companies are assigned to search operations on other floors as requested. The fifth truck is the FAST crew.

continued on page 59

Above: Paterson, New Jersey, fire companies en route to an alarm called their dispatch to report a large header emanating from a different location. Other companies arrived to find heavy fire in a lumberyard. The fire threatened multiple exposures, and the location required a long hose lay to get a steady water supply. Additional alarms were sounded, bringing mutual aid to fight this fire. Paterson Ladder 3, an E-ONE 100-foot TDA, is being used to deploy an elevated master stream. *Michael J. Coppola, PublicSafetyPictures.com*

Left: In the early 1990s, the Baltimore City Fire Department began updating its fleet with Pierce Saber pumpers. A 2000 model was assigned to Engine 85. These rigs marked a change for Baltimore to a red-and-white color scheme from the white-and-orange design they had used for many years. The Saber chassis was a lower-cost offering from Pierce and was popular with many departments that were unable to buy the most expensive units.

Above: Firefighters in Buffalo, New York, pack up two engines after a fire in a commercial building. Hose must be repacked at the scene for proper deployment at the next fire, whether it occurs in a matter of minutes or days. The solid red rig in the street represents an old color scheme. The newer fleet is painted white over red.

Left: Many fire departments spend the bulk of their time responding to calls for EMS. Here, firefighters from Station 26 attend to a stabbing victim at a public housing project. They responded with an engine, a heavy rescue, a medic unit, and an EMS supervisor, along with a sizable group of county police officers who ensured that the scene was safe for the firefighters. On their belts, these firefighters carry small packs with barrier protection devices to protect them against communicable diseases.

continued from page 56

The rescue and squad companies will assist with rescues, searches, or any other assignment deemed necessary by the IC. The rescue operations battalion chief will supervise the rescue and squad companies if they are engaged in rescue work, or may be used for firefighting operations.

San Francisco

In San Francisco, the San Francisco Fire Department (SFFD) staffs engines and rescues with one and three—one officer and three firefighters. Ladders are staffed with one and four. Every truck company in the city has a 100-foot LTI tractor-drawn aerial (TDA). A full box-alarm assignment consisting of three engines, two ladders, one rescue, one ambulance (with one firefighter paramedic and one firefighter EMT), two battalion chiefs, and an assistant chief with a driver/aide are dispatched to a reported fire in a high-rise. Any working fire gets a fourth engine as an RIC and a paramedic supervisor. A working fire in a high-rise automatically is upgraded to a second alarm, which gets four more engines, two more ladders, two additional chiefs, a rescue captain, the mobile air unit, an arson investigator, and a representative from the bureau of equipment.

The first and second due engine companies, the first due truck, and the first due battalion chief are assigned to the fire attack team. One of the box-alarm engines is responsible for lobby and elevator control, and the other establishes a water supply with a manifold in front of the building. The rescue squad is assigned to search the fire floor and the floor above.

The SFFD has preplanned alarm responses from the first to the fifth alarm levels. After that, requests for additional resources are special calls.

Buffalo, New York

In Buffalo, a high-rise box assignment is sent to all reports of fire in a building with eight or more floors. The response includes four engines, three ladders, one heavy rescue, one air truck with spare SCBA bottles, a division chief, two battalion chiefs, and an EMS supervisor. A second alarm will be requested immediately if there is fire or smoke showing on arrival. Buffalo has minimum manning of four on all engines and ladders; the heavy rescue has a crew of five. They get 38 people on the original dispatch.

The first engine company and the first ladder company will go to the fire floor. The engine will put the first line in service, and the ladder company will perform searches, rescues, or forcible entry, or it will assist the engine company as needed. The second engine company is responsible to charge the standpipe system and secure a hydrant. The crew will await orders from the IC. The third engine also awaits assignment from the IC, along with the second ladder company, while the fourth engine will assume lobby control. The third ladder on the alarm will be the FAST. The first arriving battalion chief establishes command until the division chief arrives and assumes forward command. The EMS supervisor may initially assume the position of lobby control, as could the driver of the first suppression company on the scene, depending on who arrives first.

A second-alarm high-rise assignment in Buffalo brings an additional four engines, three ladders, and two battalion chiefs.

Fairfax County

The Fairfax County Fire and Rescue Department sends a first-alarm assignment of four engines, two trucks, one rescue squad, one EMS unit, two battalion chiefs, and two EMS supervisors to a reported residential high-rise fire. They staff engines with four, trucks with three, and EMS units with two, and the rescue squad has three firefighters and an officer. Along with the battalion chiefs and EMS supervisors, this puts 32 people on the scene.

Assignments for the first-alarm companies stipulate that the first two engines and the first truck go *continued on page 62*

Chicago firefighters use the master stream from their tower ladder to hit a fire on the top floor and roof of a five-story building. Only after interior companies are out of the building can these lines go to work. The stream from another line originating on the lower roof of a building across the street can be seen above the heads of these firefighters.

Chapter Two

Right: Sometimes the hardest work is performed after the fire is out. Overhaul is very labor-intensive and can be quite involved, depending on the circumstances of the fire and the extent of damage. Here a firefighter uses a Halligan bar to pry the fascia off a section of the building exterior to check for hidden fire extension as water cascades down upon him from the floor above.

Opposite: Chicago runs four heavy rescue squads. Three are in the city, while the fourth is at O'Hare International Airport. Each of the city squads runs as a two-rig company consisting of a 55-Snorkel and a more conventional heavy rescue, both on HME chassis. Squad 2, which covers the north side of the city, positioned its Snorkel right in front of this 3–11 alarm fire on Devon Avenue. Two other elevated master streams are visible working farther down the street. The short wheelbase and 55-foot articulating device give the squads excellent maneuverability and access to get right into this type of fire and get to work quickly.

Below: The maneuverability of the Snorkel allows firefighters to alternate between directing their stream over the roof to a position just above street level to get into the second-floor windows. The jack spread of the stabilizers on the Snorkel is also considerably smaller than that of the tower ladder down the street. The beauty of a large-city fleet is that it is able to take advantage of multiple types of apparatus to fit many needs, as compared to a small department that must decide which single unit will best handle their requirements.

This is Chicago's large deluge gun. Unit 671 is roughly twice the size of the smaller units and is capable of discharging a greater amount of water and directing it farther. A maximum of ten 3-inch lines can be connected to this rig for a couple of very powerful streams. Like the smaller units, an engine company will be taken out of service and detailed to bring this rig to a scene if it is requested.

Many seasoned firefighters in areas that have cold winters actually prefer the cold to the heat and humidity of summer. In cold temperatures, they can bundle up and take advantage of the heavy layers that make up their bunker gear. During the hot summer months, they still wear the bunker gear, which limits the time that they can function without the threat of heat exhaustion. This firefighter is taking a break from operating a master stream in the bucket of an aerial. Due to the extreme cold temperatures, the water freezes on every surface.

continued from page 59
to the fire floor. The third engine takes control of the lobby until the battalion chief arrives. The fourth engine and the second truck go to the floor above the fire, and the rescue squad establishes a RIT one floor below the fire. The first battalion chief and EMS supervisor establish command in the lobby, and the second battalion chief and EMS supervisor report to the fire floor and become the fire attack division.

If the incident requires additional resources, a second-alarm assignment would be dispatched bringing three more engines, one more truck, one EMS unit, a battalion chief, another EMS supervisor, a light and air utility vehicle, a safety officer, the deputy chief of operations with an aide, a field command unit, and a canteen. The field command unit comes with a crew of four, and two come on both the canteen and the light and air unit.

Los Angeles City

The Los Angeles City Fire Department (LAFD) runs engines, light forces, task forces, squads, heavy rescues, and urban search and rescue (US&R) companies in addition to other specialized units. All of their truck companies are 100-foot TDAs and are assigned to a light force or task force. The light force is a ladder and an engine staffed with an officer and five firefighters. Four ride on the ladder and one firefighter
continued on page 65

Firefighters from Los Angeles City and Los Angeles County fight a fire at a recycling center, where bales of cardboard and newsprint produced 50-foot flames. Firefighters are preparing to put Ladder 38 to use deploying an elevated master stream to supplement the hand lines. Los Angeles and several other fire departments in Southern California wear helmets that are distinctive to that region. *Jon Androwski*

EMERGENCY MEDICAL SERVICES

Many fire departments across the country have incorporated emergency medical services (EMS) into the services that they provide for their district. If the fire department does not handle EMS, then it is possible that the community uses a private contractor, a separate rescue squad, or perhaps another division run by the local department of health. Regardless, in most parts of the country, almost any call for EMS will involve the fire department. Firefighters are often trained as emergency medical technicians (EMTs) or the more advanced level of paramedics. Many fire departments have advanced life support (ALS) engine companies that are staffed with paramedics and carry all of the specialized equipment that is carried on an ALS ambulance, but the engine does not have the means to transport a patient. A fire suppression company is dispatched with the EMS unit in most areas of the country. The firefighters may be able to arrive sooner than the ambulance, since there are generally more fire suppression companies on the streets than ambulances. In addition, the busiest ambulances spend much of their shifts at emergency calls or transporting patients to the hospitals.

Since the number of fires is down for many fire departments, EMS calls account for the majority of their responses. The largest cities were probably the last to incorporate EMS into the fire department, although in many cases the personnel that respond on an ambulance are not firefighters but a separate division under the auspices of the fire department. Other cities rotate personnel between fire companies and medical units. Citizens often wonder why a fire truck with three to five firefighters shows up at a call for medical help. Aside from the fact that the fire company may be able to arrive sooner, there is often a need for extra hands to help carry equipment, lift a patient, or maneuver the stretcher out of the building.

EMS duties span a wide range. They include responding to medical emergencies, motor-vehicle accidents, emergencies requiring life-support services, calls for assistance from people who cannot help themselves after a fall, and calls to provide medical assistance to firefighters at the scene of a fire. An EMS unit routinely stands by at a fire scene in the event that a firefighter is injured. The EMS unit at a fire scene also serves as the rehabilitation (rehab) sector, where firefighters are sent after they are replaced with fresh crews. In rehab they cool down or warm up, depending on the weather, and they can get water or hot drinks from a canteen or other support unit, which supplements the EMS unit. Firefighters may also have their blood pressure monitored.

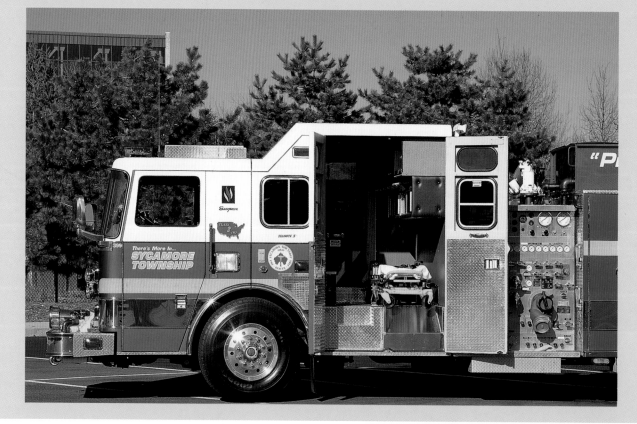

Many fire departments provide EMS. Since the number of requests for EMS is so high, fire companies respond with the ambulance and may arrive first on-scene. Often, engines are staffed with paramedics, making them an ALS or paramedic engine company. With all of the ALS engines in service throughout the country, the Seagrave seen here, one of two twins from Sycamore Township, Ohio, is a rarity. The cab was configured to accommodate a stretcher to provide transport capabilities in the event that all ambulances are tied up.

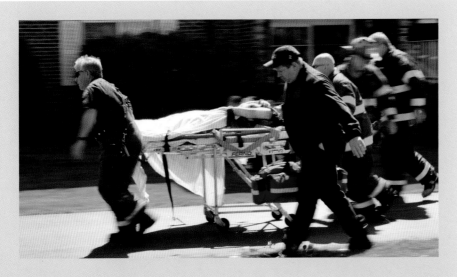

Firefighters and paramedics rush a victim from an apartment fire to a waiting ambulance. The firefighters assigned to the ambulance will be assigned to fire or EMS duties depending on the specific requirements of the incident. Some fire departments do not run the EMS or do not staff ambulances with firefighters. In these cases, the EMS personnel will stand by strictly as a medical sector and will not be involved in firefighting duties.

After extricating a seriously injured victim from a car, firefighters who are also paramedics transfer the patient to a medivac for transportation to a trauma center. The helicopter generally has a staff of three, including the pilot. This form of transport is common in rural or remote areas but is also used in urban areas, since several factors can delay the time it Atakes for a ground ambulance to get to the hospital. These factors include the time it takes to extricate a victim, traffic levels, and the distance to the trauma center or burn unit. Some fire departments that are within a reasonable distance to the trauma center feel that the tradeoff of driving time versus the time it takes to transfer the patient to the medivac and for the helicopter crew to acclimate to the patient is not worth the potential risks involved with utilizing a helicopter.

continued from page 62

drives the engine. A task force is a total company of ten, with another engine that is staffed by four. The light force engine is referred to as the pump when the company responds as a full task force. The squads also have a company of four. Firefighters assigned to a light force or a task force are cross-trained to perform engine or truck work. This allows a light force to put a hose line into service from their pump to attack a fire if they are the first company on-scene. Engines also carry forcible-entry tools.

A first-alarm assignment for a reported structure fire in a high-rise consists of four engines, two light forces, one battalion chief, one squad, one EMS captain, one basic life support (BLS) rescue, and one advanced life support (ALS) rescue. An airborne element is also part of this response, which includes one task force and two helicopters.

The LAFD has six initial objectives at the incident: fire attack, lobby control, staging, search and rescue, evacuation, and establishing a base. At a non–high-rise incident, secondary companies are sent to a staging area, which is close by but not at the incident. The units stage there until they are summoned to move up to the scene and given an assignment. At a high-rise fire, this area is called a base because staging is set up two floors below the fire floor.

The LAFD has standard operating procedures for high-rise fires, but it deploys resources based on the immediate objectives at hand. Generally speaking, the first-in company is assigned to fire attack. This could be an engine or a light force. The second company handles lobby control, and the third company identifies a suitable area for staging two floors below the fire and then can be assigned to back up the fire attack crew.

After this initial response, it is up to the IC to determine additional resources that are needed at the scene and request them from the operations control dispatch (OCD).

Anything above a first-alarm assignment automatically gets a RIC response, which is made up of one task force, one battalion chief, an ALS rescue, and a US&R company.

Subsequent to the Cook County Administration Building fire in Chicago, the LAFD immediately initiates a search and evacuation component for any trapped occupants.

Denver

In Denver, the report of a fire in a high-rise building initiates a task force assignment, which is an upgraded response from a regular first alarm. They

Above: The officer on FDNY Rescue 1 and his company stand by for assignment with other companies. In addition to his PPE and hand tools, he carries a TIC in the event the company is assigned to search a building for victims.

Opposite: Three tower ladders are working at what is left of a block-long commercial complex after a 5–11 alarm fire on the northwest side of Chicago. The 5–11 designation stems from the days before radio communication, when every firehouse had a joker stand, which consisted of a telegraph key, a telegraph sounder, and a register, which was like a ticker tape machine with a bell. The key was used by companies to send a numeric signal to the alarm office to communicate their status. The alarm office would transmit box numbers and alarms a total of four times over the register, two times each on a different circuit for redundancy. They would tap out a message like a Western Union Telegraph, and a series of marks would stamp out on the tape corresponding to the box locations; at the same time, the bell would indicate the code. Firefighters would then check the running board that was printed with each box in their nearby response district or pull the appropriate box card from a file, which would dictate the companies that were due on extra alarms. Each firehouse had a dead-man switch, which had to be held down during the last two signal transmissions, or all of the bells in the house would sound. In the 5–11 designation, the first number would be tapped representing the alarm level followed by the 11 more taps to signify an extra alarm. Hence, the 5–11 was a fifth alarm fire. Chicago is a tradition-rich department and continues to use many of these original designations and company signatures every day.

send four engines, two trucks, one heavy rescue, one hazmat company, and two district chiefs. All fire suppression companies in Denver are staffed with a minimum of four, so this alarm brings 34 firefighters and officers to the scene. Denver sends a fire attack group, which consists of the first two engines and the first truck, to the fire floor. The driver/engineer of the first engine makes the standpipe connection outside the building, and the driver/engineer of the second engine gets a hydrant and feeds the first engine. The first truck officer remains in the lobby as lobby control/systems to maintain firefighter and company accountability, recall the elevators, and make public address announcements to the tenants. As the fire attack group assembles one floor below the fire to put a line to work on the fire floor, the engine officers familiarize themselves with the floor plan, and the truck company searches three to five floors above the fire for victims in the attack stairwell.

The second truck will go to the floor above the fire to perform searches. The third and fourth engines stage on the floor below the fire and will put a second attack line to work or will take a line to the floor above the fire, depending on the situation. The heavy rescue company goes to the top floor to perform reconnaissance, search for victims, check smoke conditions, and inspect the attack stairwell.

The first-due chief assumes command outside the building, and the second-due chief will be the fire attack chief on the fire floor.

Denver upgrades to a second alarm immediately if the task force assignment finds a fire on arrival. This brings three more engines, two more trucks, and two more chief officers. One chief is assigned to lobby logistics and the other to searches and evacuations.

Chicago

The Chicago Fire Department responds to a still alarm for the report of smoke or fire in a high-rise with four engines, four trucks, one two-piece squad, three battalion chiefs, one ALS ambulance, an EMS field officer, and a mobile communications unit. Staffing is 5 each on the engines and trucks and 6 with the squad for a total of 52 people. In Chicago, a Still Alarm initially was a silent alarm that was not transmitted to the firehouse over the telegraph but was reported in person. Today, the Still Alarm is a bit more complicated but is a reported fire that is not received through a formal alarm system. It is an initial response that is less than the response for a confirmed fire. (In some cities a Still Alarm is a single-engine response.) If smoke or fire is confirmed, the incident will be upgraded to a high-rise box alarm, which

Above: What's not to love about this job? Perched high above a raging fire directing a master stream from the Snorkel's bucket for hours on end is a dream for many and reality for only a few. This firefighter from Chicago's Squad 2 was busy for hours before he went down for relief.

Right: As fire vents out a bedroom window, firefighters move in quickly with a hose line but exercise caution before flowing any water because they are aware that companies are operating inside the building. Once they confirm that their actions will not endanger the interior companies, they can knock down the remaining fire.

brings four additional engines, four more trucks, one deputy district chief, one helicopter with a pilot and a battalion chief, and one SCBA support unit. Additionally, Chicago would also initiate an EMS Plan 1, which is a first alarm for ambulances. The Plan 1 gets a response of five ALS ambulances, four battalion chiefs, two EMS field officers, and one EMS assistant deputy chief paramedic. An additional 61 people are represented with the box alarm. The Still and Box Alarm is an upgrade for a confirmed fire. The Fire Alarm Office can upgrade to a Still and Box Alarm before units are on the scene based on multiple reports of a fire. Originally, the Box Alarm was received over the telegraph relaying a particular street box or pull station that was activated.

The first battalion chief acts as the IC in the building lobby. The second chief is the high-rise chief and establishes the forward fire command two floors below the fire floor. The third chief is the fire attack chief (FAC) and takes command on the fire floor.

On the still-alarm response, the first engine is the fire investigation team (FIT), which is responsible for the first attack line. They bring 200 feet of 2½-inch line with them, initiated from one floor below the fire floor. The second engine company assists with the first line. The third engine company is assigned to lobby control, and the fourth engine either initiates a second line or provides relief for firefighters on the first line.

The still-alarm truck company assignments are as follows: The first truck is part of the FIT and is responsible for conducting primary searches on the fire floor. This truck also assigns one firefighter to the fire communications panel in the lobby and another to operate the elevator for fire department personnel. The FIT may also have the responsibility of determining the evacuation and attack stairwells. The second truck company also conducts searches on the fire floor. This company may also be assigned ventilation duties. The third truck is the rapid ascent team (RAT) and is responsible for going to the top floor to initiate a search from the top of the attack stairwell to the bottom. They may also be assigned ventilation duties in the stairwells. The fourth truck is also assigned as part of the RAT to search the evacuation stairwell from the top floor down to the ground. This company may also be assigned to conduct a search of

Left: Big fires require enormous resources. Rigs, manpower, equipment, hose, and water are all necessary components of a successful attack. Once it's all over, everything needs to be picked up and returned to the way it was before so that all companies get back into service for the next alarm. Engines carry hose, as do some trucks. Large-diameter hose is used to supply water from a hydrant to a rig or between rigs. As fires grow, it becomes necessary to manage the source of the water to protect the pressure going to the engines. Eventually, firefighters have to be concerned about how much water they are drawing from any given water main and may need to go farther to tap a different line. This often means that engines have to lead out more hose. Since hose beds on each engine are meant to balance the needs of hose for supply and attack, supplemental hose might be needed from other engines or hose wagons. A hose wagon is most often a rig whose sole purpose is to carry a huge amount of large-diameter hose for those extra-long lead outs. Here, at the equivalent of a seven-alarm fire in suburban Alsip, Illinois, a neighbor to Chicago, additional resources were requested from the city. In this image, several companies of Chicago firefighters work together to repack the hose from one of Chicago's hose wagons that carries approximately 4,000 feet of 5-inch-diameter hose. The rubber mats visible lying over the tail board and into the street are meant to serve as a buffer between the metal bumper and the metal couplings as the hose plays out behind the wagon as it is driving.

The Chicago Fire Department was the first to utilize a new product for use in the fire service in 1958. This product would later be called the Snorkel, and a long-standing relationship between the two still exists today. Although the only front-line Snorkels still in service in

Chicago are the three 55-foot units assigned to the squad companies, this larger Snorkel has been relegated to reserve or special-call status. It is not uncommon for the IC to call this Snorkel to the scene of a large extra-alarm fire to be deployed as an elevated master stream. In this scenario, an engine company will be detailed to go out of service to bring the Snorkel to the scene and handle its operation. Reserve Snorkel 1 is pictured here working during the late stages of a 3–11 alarm fire on the city's southwest side. The below-freezing temperatures cause water to freeze on all surfaces, as illustrated by the tree branches.

the floor above the fire or conduct stairwell ventilation. The squad will be assigned to perform searches above the fire floor, to separate into two teams to assist the RAT in each stairwell, or to perform ventilation duties as directed by the IC or the FAC.

The box-alarm engines are assigned duties as necessary by the IC, which may involve relief, deployment of a second line to the fire floor or to a floor above, or search duties. Box-alarm trucks will receive assignments of relief, RAT searches, floor searches, ventilation, and/or large-scale evacuations.

One box-alarm truck will be designated as the RIT company. One of the box-alarm battalion chiefs will be assigned as the RIT chief. The deputy district chief becomes the IC and determines the need for additional resources.

Houston
The Houston Fire Department outlines four tactical priorities for personnel at high-rise fires. They are the rescue of occupants, protection of exposures, confinement of the fire, and extinguishing of the

FDNY firefighters battle a fifth-alarm fire in a "taxpayer" in the Inwood section of Manhattan. Upon arrival, firefighters encountered heavy fire involving multiple stores. Here, elevated master streams from five tower ladders—plus several hand lines—are being deployed into the building. Firefighters can be seen on the roof of an adjacent building deploying another line. *Matt Daly*

Long after the dramatic fire and smoke are knocked down, firefighters continue to lob large quantities of water on the ruins to douse what may now be a smoldering or burning pile of rubble. One 3-inch line is tied into this portable monitor and watched by a firefighter. It will likely be in this position for quite some time.

To aide in combating large defensive fires, the Chicago Fire Department has three deluge wagons on the roster. Two of them are on one-ton pickup chassis like this one with dual guns; the rigs have multiple large inlets on the tail that feed each gun. These vehicles are technically in reserve status. Although many years ago they were due on certain alarms, today they must be special called to get them to a scene. An engine company will be detailed to bring one out, set it up, and operate it at an incident, which in turn places that engine company out of service as an engine.

fire. They acknowledge that there are times when the best way to protect the building occupants is to extinguish the fire quickly while occupants remain in their apartments or offices. At the same time, they stress the importance of confining the fire to the original floor and for officers to ensure that they have sufficient personnel and resources in place prior to initiating an attack.

Staffing for fire-suppression rigs is a crew of four. The first-arriving engine has to investigate the fire and upgrade immediately to a second alarm upon confirming a fire. This officer becomes the initial IC until the arrival of the first district chief. The engine company becomes lobby control and operates the first elevator for other companies. The first ladder company, along with the second and third engine companies,

becomes the fire sector team (FST) led by the ladder officer. They proceed to a floor below the fire to prepare to begin a fire attack. Prior to initiating an attack or rescues, the FST determines the attack stairwell and evaluates a site below the fire floor for the resource pool, a staging area for personnel and equipment.

The second and third ladder companies plus the fourth and fifth engine companies are given assignments by the IC based on the most immediate needs, which could include a backup line, searches, rescues, or RIT. The engine operators from the third, fourth, and fifth engine companies work together to ensure that the standpipe is supplied by one or more engines. The second district chief assumes control of the lobby, and the third district chief receives his or her assignment from the IC.

Philadelphia Engine 50 is at the hydrant pumping in-line to a second engine down the block. The KME rig has a front intake so the driver can pull right up to the hydrant. The rig down the block is using a deck gun.

CHAPTER THREE

RURAL FIREFIGHTING

Rural firefighting is most often associated with volunteer firefighting. Seventy-two percent of all firefighters in the United States are volunteers. It goes without saying that the more remote, less-populated areas of the country have less fire duty and fewer total calls for help than more-populated areas, but there are, nonetheless, lives and property to protect. This responsibility rests on the shoulders of dedicated and outgoing individuals who are willing to spend time, make commitments, and sacrifice themselves when they are called upon to come to the aid of neighbors and complete strangers alike. Although most rural fire departments are made up of volunteers, it is not the case that all volunteer firefighters are in rural areas, nor is it the case that all rural firefighters are volunteers.

In years past, volunteering was different. Firefighters could receive a minimum or a more advanced level of training depending on their level of commitment. They could come to a few regular meetings and then be prepared to respond to calls for assistance. As the fire service has progressed, nationally recognized standards have been implemented across all levels of the fire service with regard to training, firefighter safety, on-scene accountability, rapid intervention, and mass casualty or disaster preparedness. The expectations and requirements of volunteering in most areas have increased dramatically. This added training and the associated commitment of time that is required to complete it, coupled with the constraints of a full-time job in addition to family obligations, equates to a growing problem in filling the ranks of volunteer fire departments. It has also become a struggle to get sufficient numbers of people to respond to an emergency. Many firefighters who volunteer near their homes may commute to a job that is a long distance from the fire department, taking them totally out of the loop when needed to respond to an incident during their work hours. In addition, many people simply are not interested in working for free.

STATUS QUO

As with any generalization, there are exceptions. For example, rural fire departments in some of the more remote areas of a few states remain largely as they have been for decades. They are generally 15–20 years behind the fire service in terms of training and testing of personnel and equipment. That is to say that they do meet regularly for training at the firehouse, but this dubious training may consist of card playing, eating, drinking, and socializing. There is a mindset in some places that they've been fighting fires the same way for many years, and they have the experience needed to continue doing so. This involves lobbing water on the fire until it is extinguished. These departments have little external oversight and consist of residents who are content with maintaining things the way they've always been. Fire chiefs, in these instances, usually maintain their leadership positions until they retire or quit. Many of the volunteers are friends or family members. It is not uncommon for children to follow siblings and parents into the fire service. Upgrades in equipment, tools, and personal protective gear largely come in the form of hand-me-downs or donations from other departments that have upgraded their own goods. In contrast, some rural areas, however, have full-time career coverage ranging from a single driver to a full company of firefighters.

Opposite: Firefighters monitor the progress of a tender dumping its water into one of two portable tanks that have been set up together to allow this engine to supply rigs closer to the fire. The tank closest to the engine appears almost empty. A 3,000-gallon portable tank can drain quickly if it is supplying multiple hand lines or a master stream flowing several hundred gallons of water per minute.

A firefighter opens the rear chute of this elliptical tanker/tender with an electronic switch, which differs from the side-dump chutes that have mechanical levers. The vertical racks on either side of the tank are storage racks for the portable folding tanks.

Below right: Firefighters, whether they are career or volunteer, all look the same once they are wearing bunker gear. This image presents an interesting representation of three stages of dress for the Secaucus, New Jersey, volunteer department. One firefighter, hardly visible at the pump panel, is in street clothes, while the firefighter who is with him is in full bunker gear. The third firefighter pulling the red line is half in and half out. Secaucus Engine 5, shown here, is an older engine built by FMC, which no longer has a division building fire apparatus. The cab and chassis is a Ford C-8000 series, once a staple of the fire service, with an open canopy crew section that was added by FMC. This April 2006 fire, which eventually went to a third alarm, was in the Meadowlands along the New Jersey Turnpike. Initial reports were for a small brush fire with light smoke conditions on arrival. Access to the incident was difficult, and the lack of water on the Turnpike resulted in the fire's rapid growth. Heavy smoke conditions forced portions of the Turnpike to be closed for several hours. *Ted Pendergast*

CHALLENGES

Daytime has always been more difficult for volunteers to respond than the night because many work their full-time jobs during the day. It has become harder and harder for many workers to have the flexibility to leave work in order to respond to an emergency. When calls for the fire department are for automatic fire alarms or other non-confirmed emergencies, leaving work can be difficult. Rural departments are increasingly relying on mutual and automatic aid from neighboring departments to help ensure that a sufficient number of people respond to a scene, enabling both the efficient and safe mitigation of the emergency. If each department is able to get at least one rig on the road, they stand a better chance of not being overwhelmed on-scene. Adding to the difficulty is the fact that many rural fire stations are spread far apart, and districts can be enormous. It becomes easy to understand the plight and difficulties facing rural fire departments throughout the United States.

Rural areas are vast, often with long two-lane roads that twist, turn, rise, and fall as they traverse hilly terrain. There are often narrow shoulders and long stretches between structures, homes, or towns. Police patrols in rural areas are scarce, and the roads are not illuminated. Drivers have a tendency to travel at high speeds on these roads, which creates a dangerous scenario and a higher-than-normal risk of motor-vehicle accidents. All of these elements combined lead to some horrific wrecks out in the middle of nowhere, far from the police, fire, rescue squads, ambulances, hospitals, and perhaps even someone to come upon the scene to call for help. Severe single-vehicle accidents are common in rural areas, as are multi-vehicle collisions, either at crossings or involving head-on crashes.

All of this leads to some interesting conclusions about the rural fire service. Whereas municipal firefighters may receive extensive training and have a career in the fire service, many rural firefighters come into contact with more severe accidents to hone their skills on the job. Considering that a career firefighter works on a rotating shift, and a volunteer firefighter is always on call, there is a higher likelihood that the volunteer will respond to more serious wrecks, which require a high degree of skill to rescue a victim or victims. Volunteers must be proficient with their skills because there may not be any backup. More often than not, they will be requesting medical helicopter transportation for their patients due to the remote locations of the wrecks and the long distance to a trauma center or hospital. Unlike the municipal firefighters, volunteers will often have to make do with

The Lower Mimbres Fire Department in Faywood, New Mexico, is basically in the middle of nowhere. Nevertheless, the department has a dedicated group of firefighters who manage their own ranches and give of themselves to help others. They are in such a remote area that waits of 20 to 30 minutes for ambulance transport aren't uncommon. If it is possible, first responders take patients in a vehicle to try to reduce the distance that the ambulance has to travel. The department recently purchased two commercial pumpers from E-ONE on Freightliner chassis. The rigs are not fancy and have no frills, but they fit the bill to respond to the needs of the community.

Right: Rural volunteer firefighters must respond to an incident from home or work when they are summoned. Many carry their PPE with them in their own personal vehicles so that they can respond directly to the scene, while others bring apparatus. Sometimes it's hard to tell who's on the fire department and who's a bystander. Five firefighters responded with two pieces of apparatus to this collision in the western portion of central Indiana.

Below: When victims are seriously injured far from a trauma center or burn unit, they are routinely transported via helicopter medivac. The fire department requests the helicopter as soon as they determine the need for air transport. The department assigns units to establish a landing zone either at a predetermined site or at an appropriate location near the scene. This victim of a motorcycle accident was not wearing a helmet and sustained serious injuries. The landing zone was established in the parking lot of a small strip mall, where the ambulance brought the patient to meet the medivac. As firefighters ensure a safety zone for civilian bystanders, others help the flight crew load the patient into the helicopter.

fewer personnel for the same reasons stated earlier—full-time jobs and long distances to travel.

Some rural rescue squads or fire departments that offer EMS find themselves providing a training ground for nearby municipal firefighters and medics. For incidents in which ground transportation is utilized, the extended drive to a hospital or medical center provides additional time to treat patients in the ambulance. In an urban setting where the hospitals are closer, medics have only a few minutes to care for the patient before they are transferred to hospital personnel.

FIRES

It is possible that some remote areas will sacrifice the letter of the law when it comes to the national recommendations versus the ability to have personnel available to handle the emergency situations that arise in their districts. Rural fire stations can be 10 or more miles apart, where urban fire stations are generally less than 2 miles apart. This in itself increases response times, but when one factors into the equation that some volunteers have to first respond to the station to get the rigs, it is easy to see that rural areas experience some long response times. Generally, one or two firefighters will get the rigs on the road while others respond directly to the scene. Firefighters who live or work far from the station will often keep their personal gear in their vehicle. This arrangement may serve to reduce the time it takes to get on the scene of an emergency, but these volunteers may be limited as to what they can initiate without a rig. Some department policies require that a rig wait at the station until a minimum of three or four firefighters are on board. The driver will pull the rig onto the apron and wait for a large enough crew before departing. Each department has its own set of protocols.

As if these issues don't present enough obstacles to rural firefighting, fire departments have to contend with unmarked dirt roads, difficult access to isolated properties, addresses that are not marked or do not represent an easily identifiable location,

and the mere fact that incidents can go unnoticed for extended periods of time. Some counties have installed enhanced 9-1-1 systems that allow them to track the location of a person dialing 9-1-1 during an emergency. Years ago, without knowing the exact location of an incident, fire departments would have to drive around looking for the fire if they were unable to see a smoke header in the daytime sky or a glow in the dark of night. At times, all the fire department can do is protect exposures, as the initial structure is too far gone.

One common understanding in the rural areas is that a structure may burn to the ground before firefighters arrive on the scene. There is a saying that fire departments have saved many foundations. It does not take long for a fire to destroy small buildings, mobile homes, or trailers. If the fire is not detected or reported early, it is unlikely that a fire department will be able to save much of the structure, or even be able to rescue someone trapped inside. If a structure or trailer were well involved on arrival, it would not be uncommon for the firefighters to initiate a quick

continued on page 80

The Hamel Fire Protection District in Illinois purchased this pumper/tanker with a 1,000-gallon water tank to give them a head start on fires in their rural district. As shown here, they drop a portable water tank to supplement what the truck is able to carry to the scene. The commercial Freightliner chassis does not have a rear crew area for additional firefighters. The added cost of the larger cab, coupled with the likelihood that additional personnel would respond to a scene directly in their own vehicles, factored into this design choice.

There's something perhaps morbid and yet at the same time mesmerizing about watching the violent destruction caused by a fire. After the completion of training, everyone settles back to watch the fire take its toll. Even on this day where the temperatures were in the low teens, the heat coming from this fire was enough to cause discomfort some 80 feet away.

A fire in a lumberyard late at night, high winds, a less-than-adequate water supply, and very old buildings without active fire prevention devices all came together to fuel a fierce firestorm that consumed five buildings before it was finally brought under control. In the foreground are the remnants of a storage yard of the lumberyard, while the building across the parking lot in the rear lights up dramatically as a result of the high winds blowing hot embers onto the dry roof.

At the time that the fire department in Burton, Michigan, purchased two of these Pierce engines on converted four-door Kenworth chassis, their district included rural farms in addition to the more modern town areas. The Kenworth chassis allowed them to get more out of two trucks simultaneously than if they had opted for custom chassis for the rigs. This top-mount pumper/tanker has 1,500 gallons of water onboard for initiating an attack in the rural areas without hydrants.

continued from page 77
hit with a deck gun in an effort to make a fast stop. Since the NFPA mandates a minimum number of firefighters to initiate an attack, and the first-due rig may be limited by the water onboard, using the deck gun can be an attractive option.

Trailer homes are common to many rural areas. Many older trailers were built poorly and burn rapidly.

Floors in these older trailers were made of layers of pressed board, which deteriorates quickly and collapses when it is soaked with water. Another trait of older trailers is that they were constructed using inexpensive aluminum wiring. The problem that this causes stems from the fact that aluminum wire expands and stretches when it gets hot. When the wire cools, it can contract, leaving gaps that will cause electrical arcing, which in turn is a potential source of fires.

Newer trailers now have stronger, insulated glass and are sealed better, which cuts off the oxygen for fires. In some cases, this starves the fires and they snuff themselves out. Another consideration, since newer trailers are mass-produced, is that the drywall is generally glued to the walls, making it very difficult to pull

On-site training by experienced and accredited instructors is vital for suburban and rural fire departments alike. Here, after completing a training evolution, instructors go over the lessons learned, what went right, and also what went wrong. No one gets singled out or takes anything personally, with the exception of being concerned that they might not have done all they could have to support the other firefighters. It's all about learning and gaining knowledge from those with more experience.

This shot depicts operations at a house fire with a tender shuttle. The E-ONE Quint is being prepared to deploy an elevated master stream, which will require a consistent water supply. As the firefighter at the tip gets situated to go to work, companies are waiting for the lines to be charged from the engine at the top of the road.

CAREER FIREFIGHTERS WHO VOLUNTEER: WORKING BOTH SIDES

Some firefighters can't get enough of the job. Many volunteer at other fire departments on their off days. Some do this because they feel the need to give time to their communities, while others began volunteering before they were hired on by a career department. Regardless of the reason, there are many career firefighters who also give of themselves as volunteer firefighters. Career firefighters may serve as volunteer officers or chiefs. Career officers may volunteer as rank and file firefighters. Career chiefs may hold similar positions as volunteers. Either way, the commitment is the same, although there are similarities and differences in the jobs and cultures.

There are big differences between the two positions. The career department offers a family atmosphere and attitude. Working, eating, training, living, and putting your life in the hands of the same group of individuals day in and day out over a lengthy career create bonds that are sometimes stronger than the family bonds at home. Volunteer departments, on the other hand, train together, respond to calls together, and may socialize together, but the bottom line is that until they are summoned again, they all go their separate ways after the fire call ends. When the pagers or overhead siren announce the need for the fire department, the first-arriving volunteers will respond with the initial rig, and the others may come to the station, sign in to acknowledge their attendance, then sit and wait until they are released to return home. Unlike the full-time job, this may be their only contact with fellow volunteers until the next scheduled training or the next emergency. Career firefighters train while on duty. The training requirements for volunteers make it hard to find enough people who are willing to devote the necessary time to the task.

It is possible that career firefighters who have never volunteered may not have the same drive as volunteers. Countless volunteers would jump at the chance to get hired on to a full-time department and are envious of those fortunate enough to be paid to do the job. Some volunteer departments experience high rates of turnover and difficulty in maintaining sufficient numbers of personnel on the roster to handle anything other than big emergencies. Communities that no longer find themselves able to provide adequate service due to a lack of volunteers often find that the first to apply for the full-time positions are those eager few who have been regularly volunteering.

Years ago, oversight in the fire service was more lax than it is today. Going to the firehouse was a way to get out of the house. One could always find beer in the volunteer firehouses, and socializing was as large a component of volunteering as was the desire to help one's neighbor. In some ways, this is still the case; however, many fire departments have become more stringent than others.

Legal considerations affect the ability and desire for some career firefighters to volunteer. If a career firefighter is injured while volunteering, the ramifications can be great. The career department benefits may not extend to the firefighter and the firefighter's family while they are volunteering. This can jeopardize disability, death, and pension entitlements. If career firefighters have not had sufficient time at their career jobs, they may have to settle for a lesser amount from workmen's compensation. Regardless, many do not allow these factors to deter them from volunteering. Whether they do it out of a sense of duty or for the sheer love of the job, there will always be many who proudly wear two helmets.

down during the overhaul phase after a fire has been extinguished. This is tedious and exhaustive work.

Since trailers are set above the ground, they are susceptible to brush fires, which can travel underneath them. Singlewide trailers burn very quickly, especially when the fire originates from the brush underneath. Another complicating factor and fire hazard found on some farms in colder climates is actually intended by the owner as a means to provide insulation. Straw or hay from the fields is stacked around the base to block the wind and cold air. Often, this barrier is expanded and added to year after year. In addition to the obvious fact that this material is extremely flammable, fresh hay emits heat when it dries out, potentially enough to ignite the older, dryer hay alongside.

WATER SUPPLY

By the very nature of most rural areas, fire departments are required to bring water to the scene via large tankers or tenders. A fixed water supply in the form of permanent wet fire hydrants will not be available, so other means must be established to supply enough water. Large water tanks can be utilized in one of two ways. The first is called nursing, where an engine hooks up to a tanker like it would to a hydrant. The water is drawn directly from one or more tankers.

The second method involves the use of portable folding tanks that are set up on the ground to act as reservoirs. The portable tanks are stored along the side of a tanker and open easily and quickly

Four tenders in a shuttle operation wait in line to be called upon to dump their water into portable tanks being used to support a large fire attack. As each tender empties its tank, it leaves the scene and proceeds to a designated fill site to refill its tank. The tenders return to the scene and find a position at the back of the line to repeat the cycle as many times as required until the fire is out.

Rural fires often require the fire department to shuttle the water to the scene in large water tankers or tenders. The tenders dump their tanks into a portable storage tank on the ground to supply an engine, or they will hook up directly to the engine in a nursing operation. Either way, once the tender is empty, it has a predetermined fill station to restock it for another trip to the scene. In this photo, the tender draws water from a fire hydrant through two 3-inch lines. The closer the fill station is to the fire, the quicker the tender can get back with more water. Longer distances require more tenders in the shuttle.

Unincorporated areas rarely have a system of fire hydrants, an upgrade that comes from municipal government. Many areas, however, do have access to underground wells for use in an emergency. In these cases they install a dry hydrant, which is little more that a large pipe, into the well with threads at the top end that allow the fire department to hook a length of hard suction hose. Here, two firefighters are making the hookup to supply an engine at a fire scene in a commercial area.

This rig is connected to a dry hydrant, which allows the operator to draw water from a pond. The charged 5-inch line on the opposite side is supplying an engine closer to the scene, and the other line, which is not charged, snakes to an engine in the background that is drafting from a portable tank. This rig is several blocks from the fire, and the operator has little knowledge of what's happening at the scene other than what he can hear of the radio traffic.

for implementation. The tankers dump their water into the portable tank, then leave to find a source to refill. Multiple portable tanks can be set up and interconnected to increase the water that is available to fight the fire. As one tanker or tender empties and leaves, another will pull in to dump or stand by until the portable tank requires more water. The engine supplying the hose lines will drop a section of rigid, hard suction hose into the portable tank to draft the water in a manner similar to hooking up to a hydrant. As the water is pumped and the tank level drops, tankers will dump their loads before going to refill. This operation is referred to as a tanker shuttle and requires a sufficient number of tankers to ensure that the portable tank or tanks do not run out of water, endangering the firefighters manning the hose lines.

The person responsible for monitoring the tanker shuttle is the water supply officer, who has to take

into account how much water is being used to fight the fire, how large the tankers are, and how far each tanker has to travel to refill before the water is available to dump again into the portable tank. If there is a gap in the tanker shuttle, then a higher alarm will be requested for additional tankers to ensure an adequate water supply.

The fill station for the tankers can be a wet or dry fire hydrant, a standing water source, or a predesigned overhead fill line. Depending on the source or sources, an engine may be stationed at the fill site to draw water from the source and pump it into the tankers. If a pond, lake, or other standing water source is used, the filling engine may draft directly from the water source if it is physically possible to get the rig close enough. If the engine has to draft from a pond, the driver must find a safe spot to support the heavy fire engine, bringing it close enough to drop a

Two engines are parked side by side at the entrance to a street to draft from these portable tanks. They are supplying an elevated master stream and several hand lines almost a block away, which require a constant supply of water. Four portable tanks are set up to ensure an uninterrupted water source, although this in itself is no guarantee that the flow will be uninterrupted. Access on the street for tenders to maneuver is crucial, since they must be able to get to the tanks and then back out to the fill station. This fire was in a subdivision with enormous homes and quaint streets, which complicated the tanker shuttle and required an escalation to a fourth alarm to keep enough full tenders at the ready.

Two tenders dump their water simultaneously into separate folding tanks at a nighttime fire without a system of fire hydrants. Firefighters on one tender are utilizing the rear dump chute, while the other unit is dumping from the side. Tenders are constructed with dump chutes at the rear and on both sides to facilitate access to the portable tanks.

A tender stands ready to fill the portable tank when the engine needs more water. The rig drafting has a fully enclosed cab with room in the crew area for firefighters to stand. This is a particularly popular rig in the winter months, since it gives cold firefighters a spacious area to warm up. It's pretty nice in the summer, too, when the air conditioning is on. It is not unusual to see different-colored rigs at a fire with multiple fire departments.

section of rigid hose with a strainer on the end into the water. Several lengths of the rigid hose can be connected together to span a longer distance, but all of this requires more time and can be complicated by insufficient manpower, the weather, or limited visibility at night.

Another option at a pond or lake is to use a dry hydrant that is piped into the pond. This allows a more conventional type of hookup so the rig can stay on a solid surface and draw the water into the pump. The dry hydrant is a plastic pipe that is buried underground and runs into the center of a lake or pond. The other end of the pipe terminates at an area near the road or other solid ground that will easily support an engine and preferably allow the pump operator to hook up with one section of rigid hose. There is a strainer on the open end of the pipe in the lake to prevent large objects from getting in and damaging the fire pump on the engine. There is another strainer at the top of the hydrant where the engine hooks up, and the intake

on the rig also has a strainer. Prior to drafting water from the lake into the engine, the pump operator opens a valve, which, with the assistance of gravity, allows tank water to run through the hose and into the dry hydrant to flush any debris from the system before drafting water out of the pond.

REALITIES

Structure fires are less common when strong building codes and fire codes are adhered to. Better construction, the use of superior fire-resistant materials, and fire detection and protection systems reduce the number of fires starting in the first place. Some rural areas may not have building codes, there may not be the means to inspect or enforce existing codes, or people in remote areas may choose to ignore them outright. Local officials may not even know that a landowner is building a structure on a remote piece of property. If the owners and builders are not responsible
continued on page 90

Above: The folding tanks are not as light as they look. It takes several people to fold them up and store them back on the rig. If you have a little seniority, it's a great assignment to give to the young firefighters.

Below: There's nothing like the big group photo of firefighters and instructors in front of the burning building after a day of training. The shot has to be done rather quickly, though, because the heat from the fire is quite intense. The photographer holds quite a bit of power over the group for a few minutes.

Above: For a little something north of the border, the rural district in the township of North Kawartha in Apsley, Ontario, is spread over a large area. The department has three stations because the members live so far apart. They bought these three small 4x4 mini-pumpers back in 2001 to provide a number of resources to the area. Each carries only 300 gallons of water but has the ability to draft from a standing body of water. Here, one unit is drafting and nursing the second to supply the hand lines. The third unit is working off of tank water alone.

Left: A firefighter checks the gauges that monitor the pressure coming into the rig from a tanker/tender shuttle, as well as the pressure that he is sending to both attack lines. He can sense a change in the incoming supply by holding onto the 5-inch hose.

The Blue Ridge Volunteer
Fire Department in Arizona
has a beautiful district
high in the hills. The
department purchased this
elliptical tanker from Pierce
to supplement the water
they have available to fight
fires. Much of its district is
wildland or forest, and homes
are built on large wooded
lots with difficult access.

Here is another incident
where fire personnel
summoned a medivac to
transport a critically injured
patient to a trauma center.
In Maryland, the state
police have 12 helicopters
based at eight locations
around the state. The
helicopters handle law
enforcement aerial searches,
rescue missions, and
medivac services. The pilot
of Trooper 8, as this unit is
called, is assisted by a flight
medic seen here in the
black uniform.

Chapter Three

Six full tenders wait in line to be summoned to the portable tanks to dump their water. The size of the fire, the number of lines flowing, and the distance required to travel in order to refill are all considerations for the water supply officer when determining how many tenders are needed.

Below: Often tenders shuttle water to dump into a portable folding tank, but some departments use them to nurse an engine directly from the on-board tank instead of dropping an external tank. These different operations could result from the amount of water that will be needed, or perhaps the location does not allow a flat surface for the portable tank, or perhaps there is no access or space for more than one rig. Regardless, both operations get water to the supply engine, whether they draft from a tank or receive water in-line pumped from the tender.

continued from page 86
or conscientious, or are perhaps just inexperienced, they may be endangering themselves and others with inferior structures that will have a greater-than-normal susceptibility to fire.

Some rural areas are relatively poor and do not have a sufficient tax base. These areas cannot adequately support all of the government departments that work for the citizens. Some fire departments are able to raise their own money through fundraising activities, although these are generally in areas with greater wealth. A few areas cannot afford to maintain fire service without the support and funding of the local government.

An interesting story happened several years ago, when the Hendersonville Rural Fire Department in South Carolina was in danger of shutting down because they could not afford to pay the insurance premiums on their apparatus. They were funded by memberships and donations. Even though they cov-

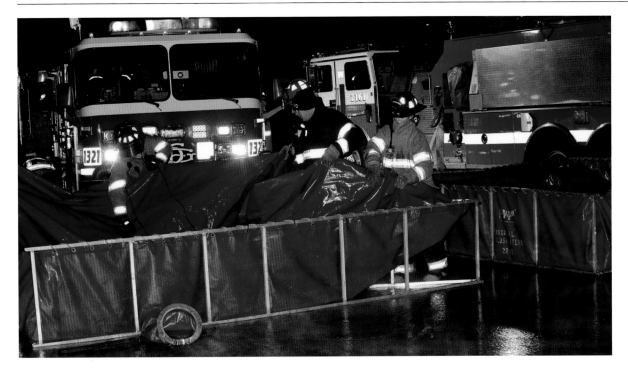

Once the fire is out and companies begin picking up, several firefighters work together to drain the remaining water from the portable folding tank prior to storing it away on the tender. Like a small backyard pool, they lift from the edges and walk toward the middle, forcing the water out through the drain.

Below: Colleton County Fire-Rescue provides fire protection for the fourth-largest county in South Carolina, covering 1,054 square miles. The department maintains an ISO Class 4 rating for the entire county and has 32 fire stations with 67 career personnel and 250 volunteers. Pictured here are four of their newest tenders built in 2007 by Seagrave on International 7400 chassis. Each carries 3,000 gallons of water and has a 450-gpm Gorman-Rupp pump. The department operates 36 tenders, 16 of which are like the units in this photo. Colleton County Fire-Rescue carries 2,500-gallon portable drop tanks in hydraulic storage racks on the passenger side of the tenders. Due to the large number of tenders in the county, they typically use their tenders to nurse other tenders or pumpers instead of utilizing a tender shuttle.

ered a large area, they were not able to generate much money, since they had few homes to protect. It just so happened that in 1993, Paramount Pictures had come to the area to film *Forrest Gump* and needed to hire a fire department to supply water to help with the simulated rain for several scenes. Another department was originally offered the job, but that chief knew of the Hendersonville situation and offered the job to the Hendersonville Rural Fire Department. This stroke of luck provided enough money (roughly $1,500) to enable the fire department to pay their bills and continue offering fire protection until the county took over fire protection about six months later. In January of 1994, all of the volunteer fire departments in the county were combined into the Colleton County Fire-Rescue Department.

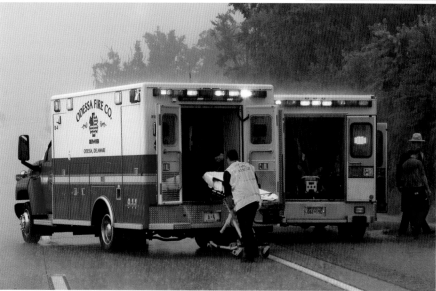

This torrential downpour made driving dangerous, resulting in a wreck on Route 1 in Delaware. Two volunteer fire departments responded and needed three ambulances to care for the victims. Firefighters from the town of Odessa were summoned by pagers and overhead sirens to respond to the station, and within a matter of minutes, several rigs were on the road.

Years ago it was common to have the pump controls at the front of a tender or rural pumper. Today this is no longer the case for several reasons, including the need for an unobstructed front grille to ensure sufficient engine cooling and the desire to keep the pump operator closer to the tools and equipment stored behind the cab.

Above: It took less than 12 minutes for this barn to collapse after the fire raged out of control. The smoke coming off this fire is fascinating to observe. It will push out from every void as the fire grows. It is a good example of the difficulty facing rural departments when called for this type of fire. It is futile to think that the structure can be saved if the fire is not caught and attacked in its incipient stage.

Right: This pumper/tender is from the Deerfield Volunteer Fire Department in Wisconsin. Engine 2 was built by Central States in 2002 on a tandem-axle HME 1871 chassis. The tall crew-cab area allows firefighters to stand while putting on their gear. The four colored lights near the top-mount pump panel are on a telescoping pole to illustrate the remaining amount of water in the large tank. When the pole is extended, it can be seen from any side of the vehicle to allow firefighters working away from the rig to roughly determine their water supply in the absence of additional rigs. This tender carries 2,500 gallons of water and 40 gallons of foam and was built with a 2,000-gpm pump.

CHAPTER FOUR

INDUSTRIAL FIREFIGHTING

Industrial firefighting differs in many ways from municipal firefighting and is most often associated with refineries, chemical plants, and power-generating facilities.

Although each of these installations has conventional structures, the primary concerns for the health and safety of the company workers, visitors, and surrounding neighborhoods lie in the areas that handle the dangerous commodities. Explosive and corrosive products in the form of liquids and gases present danger to people and the environment depending on their storage, temperature, and exposure to air, heat, or other chemicals. If a fire occurs, firefighters and plant safety personnel must have a commanding knowledge of the products and their associated dangers in order to effectively fight the fire, or in extreme cases, allow the fire to burn itself out.

Many of the chemical dangers associated with industrial installations are also routinely found within the response districts of city and suburban fire departments. These hazardous or extremely flammable substances are in locations that do not have their own industrial fire brigade, or are found in small quantities. A corner gas station is one example of an explosive and hazardous danger for municipal and rural departments. There are also risks associated with spills and accidents from the trucking and railroad industries, which routinely transport dangerous and hazardous materials. Still another example is a company with an in-house laboratory that uses chemicals as a portion of its manufacturing and may experience a leak, spill, or fire. While municipal departments have substantial degrees of training to deal with these kinds of incidents, some also rely on regional, county, or state agencies that specialize in these types of responses to provide assistance for large events.

This chapter deals with the highest-risk industrial settings that require their own on-site fire departments and/or fire brigades. Within facilities where a fire department and fire brigade exist, there is a distinction between full-time firefighters whose primary function is the fire department or public safety department at the plant, and others trained in firefighting whose full-time job at the facility can include multiple tasks. Members of the fire brigade work throughout each area of the facility and on all shifts to ensure that, in the event of a fire or other emergency, there is always someone on duty with a thorough knowledge of each process occurring within the plant. Brigade members will be the first on-scene to an emergency in their part of the facility and will begin whatever duties are appropriate before the fire department or other brigade personnel arrive. These first minutes can provide a significant head start toward mitigating an incident and saving lives. The brigade members train regularly with the full-time firefighters and work side by side during an emergency. The full-time jobs of members of the fire brigade can involve almost any specialty or position within their areas of the plant, from unit operator to maintenance.

The full-time firefighters are sometimes referred to as the plant safety department. They handle routine inspections, may take care of issuing certain permits for contractors performing dangerous jobs within the plant, and supervise the work of others where welding, drilling, or confined-space tasks are performed. Similarly, in facilities with only a fire brigade, it may be

Opposite: Here is an example of an industrial pumper built by Pierce. This engine is part of the fire department at the Valero Energy Corporation refinery in Houston. It is equipped with a Williams remote-controlled monitor, a 3,000-gpm pump, and a 1,000-gallon foam tank. The top-mount pump panel gets the operator up where he can survey the scene, in addition to allowing space for all of the large-diameter intakes on the side of the pump. Mounted on the roof is a telescoping light tower that will illuminate a nighttime scene with 6,000 watts of light.

Above: The Exxon Mobil Fire Department in Baton Rouge, Louisiana, added this custom industrial pumper from E-ONE to its fleet in 2006. Since the rig will generally respond with limited personnel, the department opted for a two-person cab that allowed for additional equipment storage behind the front seats. The engine was designed with a rear pump and a large remote-controlled Williams monitor. It also has a short wheelbase for added maneuverability within the plant.

Opposite: Fire erupts from a waste-handling facility outside Eau Claire, Wisconsin. The fire would eventually require extensive mutual aid involving 21 different departments for fire suppression companies, tenders, hazmat units, and crash trucks from the nearby airport. Local law enforcement was called upon for road closures and the evacuation of businesses. Due to the nature of the chemicals involved, a decision was made to let the fire burn to prevent the runoff of the chemicals into sewers and drains, and the potential for the contamination of ponds and streams. *Dan Reiland, Eau Claire Leader-Telegram.*

these individuals who handle inspections and monitor contractors and permits. The tasks and responsibilities vary based on the individual facilities.

REFINERIES

An oil refinery is one example of an industrial setting that is familiar to most people. Common among big refineries are large storage tanks, both cylindrical and spherical in shape; multiple vertical and horizontal vessels; several odd-looking structures of many levels with a myriad of pipes; and miles upon miles of additional piping traveling in every direction throughout the facility. The gas flame emitting from some stacks allows the excess gases from a processor to burn freely, preventing a buildup of pressure within the unit.

The cylindrical storage tanks hold liquids, such as refined gasoline, crude oil, diesel fuel, asphalt base,

Tank trucks carrying fuel and other hazards travel the highways and local roads every day. Although the incidence of fire is relatively small, the large number of vehicles increases the odds that an accident will happen. Such was the case when this tanker rolled, ruptured, and exploded on a city street. Companies can do little but hit it with large volumes of water and foam. This type of incident raises a multitude of environmental issues about the dangers of the fire itself and the environmental impact of the free-flowing product if the fire is extinguished.

heating oil, kerosene, liquid tar, or other products. The spherical tanks, often seen with a staircase winding around the outside, contain liquefied gases, such as propane or butane.

The structures, or units, within a refinery convert crude oil into different grades of gasoline or various byproducts. Some examples of these units include an alcoholization unit that makes high-grade gasoline, the crude unit that breaks down the raw crude oil, the isomerization unit that upgrades the octane of light crude, and the catalytic reformer that takes gasoline from the crude unit and upgrades the octane. Another unit on the property is the hydro-treatment unit, also known as the hydro-treater, which takes diesel and gasoline off the crude unit and strips the sulfur out. There is a sulfur plant that takes sour gas (a gas that contains too much hydrogen sulfide) from all the units and removes the sulfur, which is burned in the fuel system. Refineries also have a fluid catalytic cracking unit that heats heavy oil and cracks it or breaks the hydrocarbons into smaller particles to produce gasoline. There may also be a hydrocracker and a coker, which would break the heavy oils down further to make gasoline. Each of the units is different, which places high demands on firefighters to learn the basics about how they work, what inherent dangers each unit or processor presents, and how the tactics differ for fighting fires involving these units. Depending on the chemical in question, different precautions and safeguards must be adhered to during an emergency. For example, hydrogen fluoride and sulfuric acid are chemicals that are used in the alcoholization units. These units are surrounded by water deluge systems. Hydrogen fluoride can be deadly when absorbed into the body through skin contact and is also an inhalation hazard. Sulfuric acid will burn the skin on contact.

Comparisons

Some of the basic differences for firefighters in this arena, as compared to city, suburban, or rural departments, include the fact that industrial fires contain more hydrocarbons, which means that fires burn hotter. Municipal firefighters are trained to be very aggressive with their attack, but the industrial environment requires a more modest approach. Some chemicals within the refinery ignite upon contact with water and therefore must be attacked with foam. Temperatures in a conventional structure fire without extenuating circumstances may reach 1,400 degrees, while fires in the industrial setting can exceed 2,000 degrees when metals like titanium are involved. While municipal firefighters routinely use water as

Fire-protection systems inside a refinery or chemical plant can be used to extinguish a fire or control the spread of a fire by keeping surrounding units cool. This permanent 1,500-gpm monitor and manifold is preconnected to a foam tub. It can then be deployed to discharge foam or water depending on the need. Connected to a large, fixed, underground water supply, this manifold could be tapped to supply a fire truck or other portable monitors if an event occurred in this area of the plant.

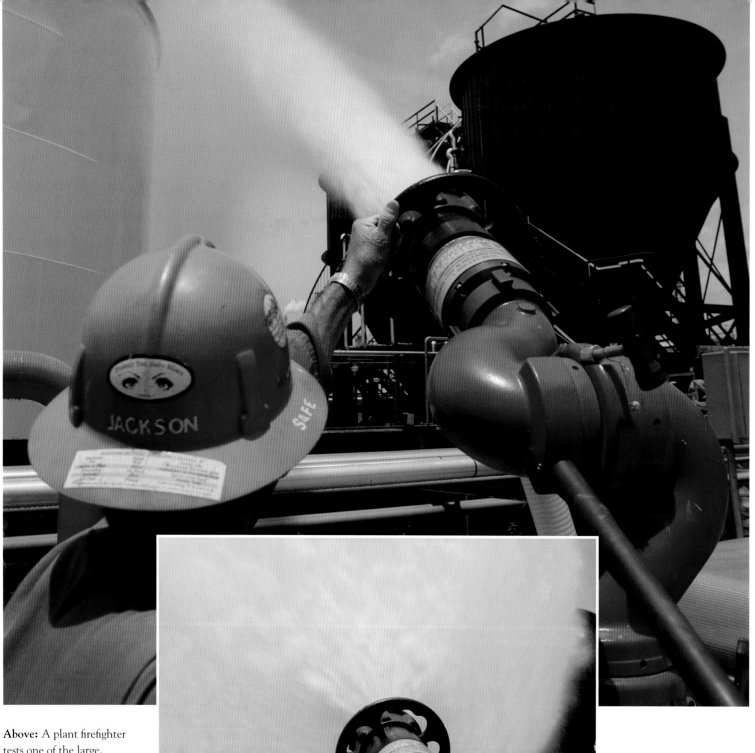

Above: A plant firefighter tests one of the large, permanent 1,500-gpm monitors that protect a series of storage tanks within a refinery. The unit is pre-piped to a foam tub to apply foam or water. Foam would be used if the tank were on fire and water would be used to keep the tank cool from a fire nearby.

Left: The monitor in the previous image is shown here with a wide fog pattern. This showers the area with a curtain of water to cool pipes in close proximity to the monitor. It also provides a cool environment for firefighters if they need to enter a hot zone to seal a leak or close a valve.

their primary suppression agent, industrial firefighters use water and foam. Municipal departments have high-rise fires in structures that allow interior access, while industrial firefighters contend with fires located in very tall towers as part of the units, which do not allow close access. Another contrast between the two types of firefighting is that conventional fires most likely are going to burn and not explode, whereas everything within the refinery is extremely flammable and often explosive. It is for many of these reasons that municipal firefighters moving into a job within an industrial department undergo extensive education and training before becoming comfortable in the new environment.

Although many materials and substances that people routinely come into contact with are flammable, the concern for explosion from an errant spark within a refinery is paramount. It is with this concern in mind that any work that produces a spark in the plant requires a hot work permit issued by the safety personnel or unit supervisors. The work must be observed by trained personnel, the qualifications of the contractors have to be verified, and the entire scope of the project is walked through before anything can begin. There is no margin for error. Some teach the premise

Repsol YPF is one of the 10 largest private oil companies in the world and operates in more than 30 countries. It purchased several units from E-ONE to protect its facilities, including this industrial pumper and trailer. The engine has a 3,500-gpm pump and a 1,000-gallon foam tank. Like any refinery, a large fire at one facility would necessitate enormous resources to flow huge volumes of water from a safe distance. The trailer offers them a secondary portable, remote-controlled monitor that they can deploy in whatever location is optimal, without being tied to multiple fixed units on the property.

that safety personnel and unit supervisors need to be 100 percent right 100 percent of the time when they are issuing permits; mistakes result in injuries.

Firefighting

Due to the extreme hazards within a refinery, the likelihood of a fire's rapid expansion, and the increased collateral danger to surrounding units, refineries are built with a great number of their fire-suppression tools permanently in place. Large, underground water supplies snake throughout the property, connecting fixed monitors capable of flowing thousands of gallons of water or foam per minute, which can be put into operation within seconds. Some monitors are preset to mix foam, while others are ready to flow only water. Portable monitors and fire trucks supplement the suppression capabilities of the fixed monitors. When a new

The Dow Chemical Company facility in Baton Rouge, Louisiana, purchased this impressive rig from Pierce in 2002. It is equipped with an 85-foot articulating Schwing boom. Schwing is known for its concrete placement booms, which were adapted for use in the fire service. This rig has a 3,000-gpm pump and carries 875 gallons of foam. The boom can be maneuvered up, over, under, or in between obstacles, including pipe racks, to hit the seat of a fire.

unit is added to the plant, engineers predetermine adequate fire suppression and design the necessary safety systems, including sprinklers and fixed monitors.

Some areas of a refinery, though, may not have fixed monitors in place. Tank farms, for example, require a large suppression capability, and the cost of installing fixed monitors to adequately cover a large tank farm may outweigh the benefits of doing so. This is one rea-

son that plants are willing to invest in large, modern fire trucks that have enormous output capabilities. Tank fires can be spectacular in scope and can burn for extended periods of time. Suppression efforts are often supplemented with mutual aid from neighboring facilities or with the help of private companies that are prepared to deploy at any time with massive amounts of foam concentrate and supplemental delivery systems anywhere in the world. Most refineries have contracts with private companies in the event of a large-scale incident, and all have mutual aid agreements in place with other facilities or fire departments.

Tank fires, the most common cause of which is lightning, can be fought in two ways: externally or internally. The external attack is with fire apparatus, fixed monitors, and supplemental monitors. Internal attacks are utilized for tanks that have been designed with a fixed foam system, which firefighters can supply from a safe distance. Fixed systems can be designed to apply foam from the top to smother the fire or from the bottom of the tank to create a barrier between the fuel and the vapor space in hopes of putting the fire out. The foam that is piped in from the bottom floats up underneath and through the product to the surface. This sub-surface attack is generally used where access from the top of the tank is limited. The top application systems are less common and may more likely appear in smaller tanks. In many cases, the cost of installing these internal systems may be hard to justify due to the rarity of tank fires. It is more cost efficient to invest in portable systems either on trailers or in trucks that can be positioned wherever the event occurs, as opposed to having to invest in fixed assets to cover every threat. Any tank fire will involve the use of spotters to view the incident from the air to help direct the operations. Fires in tanks with a 100-foot diameter require 3,000 to 4,000 gallons of foam per minute to extinguish, which is within the capability of today's modern industrial pumpers. The larger 300-foot-diameter tanks need more like 18,000 to 20,000 gallons per minute, which requires multiple monitors.

Perhaps the most dangerous tank fire scenario can occur within crude oil tanks, found mostly at oil fields. The problem occurs when water comes into contact with high heat and turns to steam. The surface temperature increases over time, but the areas below the surface are not as hot. There is a risk that some of the heavier product will sink to the bottom and that the heat will travel along with it. Water at the bottom of the tank can turn to steam as the water temperature increases, causing a violent eruption that spews hot oil out of the tank to a distance of 6 to 10 times the tank diameter.

Although fixed monitors around the plant are used for fire suppression, another extremely important task that these monitors perform involves cooling adjacent exposures to minimize the fire's extension and protect other units in the plant. Sprinklers in each unit are activated during a fire for the same reasons.

Spill fires are also a huge concern. As fuel or product spills and spreads throughout the base of the units, there is a real danger of this product igniting and involving more units and burning workers or firefighters. To mitigate this threat, firefighters deploy a blanket layer of foam on the ground to prevent ignition or re-ignition, as the case may be.

Firefighters want to minimize the units that become involved, so like any fire attack, their goal is to contain it within as small an area as possible and then extinguish it to minimize collateral damage. When fires involve hydrocarbons, water is used to cool the adjacent units, but foam must extinguish the fire. In some cases, firefighters can only contain a fire and monitor it until the fuel is gone, since greater hazards can be created by a product flowing freely if the containment vessel or structure is damaged.

Most spheres have deluge systems capable of flowing massive quantities of water, not foam. If a fire erupts in or around one of the spheres, water is required to cool the tank, since heat causes the pressurized metal

The rear of the Dow Chemical Company rig is set up to accept three 5-inch, large-diameter lines to feed the multiple master stream devices on this rig. Two fixed, remote-controlled guns are mounted at the rear in addition to the prepiped waterway on the boom.

Above: Chemical plants and refineries have multiple monitors located throughout the processing units, which can be deployed and left unattended in the event of an incident. These smaller monitors are located within the units, while the larger monitors with higher output capacities are farther away.

Above: One of the smaller 500-gpm monitors flows unattended. These units are in abundance in the more congested areas of processing units and can be deployed by any plant personnel in the event of a fire.

Opposite: Sprinklers keep pipes and fittings cool and are also concentrated in areas that are most prone to fires. The excess runoff is not wasted, since it pools on the ground to prevent the ignition of spilled fuels.

to weaken. The vapor space, not the liquid, will degrade more rapidly. Liquid petroleum gases (LPG) and their vapors are explosive and considered by some to be the most dangerous hazards at the refinery. If a pressurized container fails, there is a good probability of the vessel being launched as a projectile and putting personnel at risk. The massive quantities of water that cool the shell help to prevent these dangerous explosions.

The bottom line at a refinery or industrial institution is to prevent fires, leaks, or other disasters from happening in the first place. Job number one is prevention through inspections, preplanning, and training. During the 1970s and 1980s, fires of one kind or another were sometimes weekly events. Advances in technology, safety systems, and general plant safety initiatives have made fires a rare occurrence today. Most fires are small and are handled by the shift fire brigade members in the units prior to the arrival of the fire department or the full fire brigade.

Although fire is the most dramatic disaster at a refinery or industrial campus, there are other dangers, as well. One significant area of concern has to do with dangerous fumes. Aside from gas leaks, an

extremely hazardous environment exists in confined spaces or enclosed areas that are used for the storage or transportation of gases. Stationary tanks, railroad cars, and tank trucks require routine maintenance and cleaning. The unseen and often odorless residue of a tank's contents has the ability to quickly overcome an unprotected worker. Documented accidents describe how an unsuspecting and ill-prepared worker succumbed to the fumes while one or more other workers attempting to aid the first victim became victims themselves. To prevent this type of tragedy, one of the jobs of the safety department at a refinery entails verifying that tanks are free of gas before allowing any workers to go inside. Should a worker succumb to the fumes within a tank, rescuers must be on SCBA prior to entering the enclosure to rescue the victim. Due to tight conditions in a confined-space environment, the breathing apparatus used for this type of rescue differs in design from the SCBA used for firefighting. Depending on the access to the confined space, different methods of patient removal will be implemented, including ropes, pulleys, and other specialized rigging, to hoist the victim and rescuer to safety.

Sprinklers located throughout a refinery are used to extinguish a fire or to keep units cool to prevent additional explosions and fires. They also contain the spread of ground fires that could quickly engulf multiple areas of the plant. Most sprinklers can also be used to discharge foam in the event of a spill fire.

Here is a close-up of the monitor on top of the Exxon Mobil rig. All of the functions to operate this monitor are available to the operator from a wireless, handheld remote control.

Rigs

Apparatus owned by industrial fire departments often cover the range from brand-new, state-of-the-art units with all of the bells and whistles to 20-year-old vehicles having various amounts of factory-built and homemade attributes, and anything in between. The lifespan of a rig in an industrial department can easily exceed 20 years. It is, after all, the desire of these companies that own these plants that none of the fire trucks will ever need to be used, and for the most part, the majority of these rigs are used only for training purposes throughout their useful lives. The annual market for industrial fire trucks is relatively small, whereas the market for supplemental items, such as foam concentrate and other expendable products, is quite large. Older foam concentrates were made of animal proteins and had a useable life of 7–10 years, while newer foams are made from synthetic materials and have a shelf life of 20–25 years.

Industrial fire trucks have foam systems and can discharge very large volumes of water or foam. Engines, tankers, and aerials are all foam rigs in the industrial setting. The foam concentrate must be mixed with water. Since most foam is added to water in a 1 percent or 3 percent solution, the amount of suppression agents that these rigs can generate far exceeds the capacities of municipal fire trucks discharging water alone. Add to this the fact that the volume discharged in gallons per minute may be enormous, and the need for a fixed water supply to support continuous firefighting is no less prevalent than it is for municipal departments.

A view from the rear of the Exxon Mobil engine shows two 6-inch discharges and three intake ports. The pump panel is compact and tucked into the last compartment, allowing the operator to be close to the intake and discharge ports, which in turn do not create an obstruction to the panel.

NUCLEAR POWER PLANTS

If an architect wanted to build a facility with every possible system geared for safety, and then wanted to double those efforts with redundancies, the facility would resemble a nuclear power plant. Nuclear power plants in the United States, due to their inherent potential for a catastrophic event, contain all of the state-of-the-art automatic systems backed up by strength in building materials and further complemented by highly trained on-site personnel and fire brigades.

There are four different levels of fire brigade members. Incipient industrial brigade members are trained to take a defensive posture. They do not have PPE. They use fire extinguishers, hose lines, and ground monitors to hold a fire until others arrive. Advanced exterior industrial fire brigade members have all of the incipient training plus additional training for fires outside the buildings, including firefighting tactics for flammable liquids and gases. Interior structural industrial fire brigade members have the same training as municipal firefighters, including the use of SCBA, and possess the additional knowledge about

the systems and operations that are consistent with their facilities. An industrial fire brigade leader has further training and education to handle supervisory, command, and oversight duties.

Federal regulations require nuclear power plants to have an on-site interior structural fire brigade in order to start the reactor. The fire brigade at a nuclear plant may or may not have a fire truck. Since most of the critical facilities are in a location that ranges from a depth of 70 feet underground to a height roughly 70 feet above the ground, trucks are often ineffective. A truck is helpful, though, in providing transportation and support for emergencies in the more remote areas of the facility, which include the outbuildings and transformers. This prevents fire brigade members from having to spend the time and energy of transporting their equipment via carts to a remote section of the property. In addition, an industrial fire truck can provide large master streams for outside fires.

Inside the buildings, fire brigade members store equipment on small push carts, which are located on every floor in fenced-off areas called cages. Each

Inside a nuclear power plant, the equipment for the fire brigade is housed in fenced cages on each floor of the facility. The tools and supplies are separated into push carts that can easily be maneuvered through the facility and onto elevators if they are needed to back up the equipment on another floor. PPE is stored generically by size in lockers. Instead of having gear for each individual fire brigade member, each can grab a set of gear in his or her size, regardless of which floor each is needed on.
Bruce Boyle

attack the fire. At the same time, the plant is able to compartmentalize, or seal, every location to control the spread of fire or radiation in the event of a leak. Unless the incident is significant, others in the plant can continue with their jobs as the fire brigade works to mitigate the emergency without fear of it spreading beyond the area of origin.

Infrequently, a fire brigade will request assistance from the local fire departments. Prior to an event, several procedures occur between the fire brigade and the fire department. First of all, they train together. The fire brigade has to know that the local fire department has an understanding of the hazards involved with the power plant. Secondly, and perhaps equally as important, is that the plant has to maintain extremely high levels of security. Plant security has always been tight, though it has grown even tighter since the terrorist attacks of 9/11. Concepts and protocols have been enhanced by the Nuclear Regulatory Agency. They include such measures as upgrading the weaponry used by security personnel and implementing much more elaborate procedures for a fire department to enter the plant. Each local fire department has to submit personnel rosters to the power plant with photographs for background checks. Anyone who has not been checked out in advance is denied entrance inside the compound.

Nuclear plants do not have automatic aid agreements with local fire departments; they have to be summoned directly at the request of the fire brigade. If they are asked to respond, the fire department will be delayed at the plant entrance while security checks identification and performs a check of the apparatus. The mutual aid companies are usually sent to less-sensitive areas of the plant and are accompanied by or set up within a perimeter by security.

To reiterate a point stated earlier, the fire-protection systems in a nuclear plant are the safest of any fire-protection system. They consist of a combination of automatic CO_2 systems, sprinklers, automatic Halon fire-suppression systems, and foam systems anywhere diesel fuel is used. Throughout the facility there are hose reels connected to the standpipe system and hundreds of fire extinguishers. High-hazard areas where hydrogen gas and oil mixtures are located have automatic deluge systems that initiate with heat detectors and immediately begin flowing 1,500 gallons of water per minute. A similar system surrounds the large transformers outside the reactors. There, water acts as a coolant before the fire brigade can initiate an attack with foam to extinguish the fire. Roughly 90 percent of fires are either extinguished by the automatic systems or held in check until the fire

The pump operator of an engine company is the driver. Often this position holds a different job classification than the jump-seat firefighters. Some departments refer to this person as the engineer or the chauffer. Depending on the policies of the individual fire department, the driver can be part of the company if the rig is not pumping water, or the chauffer remains with the rig regardless of whether or not there are any lines off. This particular engine is relatively new, with a foam system and electronic station for diagnosing and monitoring all of the on-board systems. The multicolored gauges each correspond to a similarly colored preconnected line or discharge around the vehicle.

cart is able to fit inside an elevator. Multiple carts hold a variety of equipment and supplies and are stored in the uncontaminated area in the fire brigade cage. There may be carts for general firefighting with nozzles, fans, hose, and fire extinguishers; carts for foam operations with eductors for introducing foam concentrate into the water, foam concentrate, and nozzles; and SCBA carts with multiple packs and several spare bottles. Generally, a minimum of five operators per shift are dedicated to the fire brigade, working in non-essential positions that allow them to leave when the need arises. The plant has safe shutdown operators who cannot be part of the fire brigade, since they must always be dedicated to plant operations.

If a fire or smoke detector is activated, someone in the control room will dispatch the nearest operator who is on the fire brigade. This individual will investigate, or recon, the situation and determine whether others need to be deployed. If it is a matter of a false activation, the brigade member will reset or replace the detector. If there is an actual emergency, the other fire brigade members will go to the nearest cage to get PPE and the necessary cart or carts. Unlike a conventional fire department, where each individual firefighter is assigned a full set of PPE, the nuclear plant simply has a bank of lockers with common gear and helmets organized by size. This way, anyone can get a set of gear no matter which floor of the plant has an emergency. If the event is a fire, they'll hook a line to a standpipe connection and proceed to

Six area departments and a county hazmat team assisted the Croydon Township Volunteer Fire Department in Pennsylvania with a fire at a small, remote tank farm. Five tanks that were estimated to contain between 20,000 and 30,000 gallons of crude oil were involved. Firefighters fought the oil fire with foam from two hand lines and then moved in with dry chemical extinguishers for the remaining gas fire. *Jay K. Bradish, International Fire Photographers Association*

Firefighters examine the remains of a corporate jet that crashed short of a regional airport in Chicago's northwest suburbs. This is an example of a specialized fire response for suburban fire departments. The plane came down in a commercial yard, missing homes and occupied businesses, which would have resulted in greater damage and potentially more loss of life. The crew of four onboard the airplane died. The airport crash truck seen on the left was used to apply foam to the wreckage.

This is one of two vehicles that the Fairfax County Fire and Rescue Department in Virginia uses for its hazmat team. As with any large fire department with busy roads, highways, factories, and commercial businesses within its district, Fairfax County maintains a highly trained team of hazmat technicians and advisors to respond around the clock. This rig was built by Marion on a Spartan Gladiator chassis and has an interior command post, a roof-mounted telescoping light tower, and storage for many tools and supplies for use by the team. The second piece is a larger rig also on a Spartan chassis with a body built by Super Vac.

Firefighters in proximity suits demonstrate advancing on a fire with a hand line. These suits are designed to protect firefighters from exposure to extreme heat, which is most often associated with aircraft firefighting operations.

brigade arrives. The fire brigade acts as a redundancy to the automatic detection and suppression systems.

Further safeguards, or perhaps redundancies, in safety have to do with the plant's physical design and construction. Sensitive areas where an event might compromise the plant are surrounded by three-hour firewalls to prevent the spread of a fire. One example of a high-security area is the location of the pumps for cooling the reactor's core. Each redundant pump is separated and protected by three-hour firewalls with two-hour fire doors. Most of the doors have alarms and are inspected on a daily or even hourly basis to verify that they are properly sealed.

When the plant needs repair work, it may go down for a month. This process can bring 2,000–3,000 contractors into the facility. All of the work is supervised

by highly trained and technical plant personnel who can handle the critical work in any section of the plant. The plant fire marshal supervises the issuance of all permits for cutting, welding, and changes in chemicals. If work is required in the extremely sensitive areas within the three-hour firewalls, the BTUs are calculated for all flammable and burnable materials that are allowed into the area. When the amount nears the critical level that would breach the three-hour walls, nothing else is allowed inside.

The fire brigade members may also be trained as hazardous materials technicians; they are qualified to work with chemical and radioactive hazards. In the event of a radiation emergency, alarms notify everyone in the facility. Plant operations officials have minutes to account for everyone on-site. All

Left: Since the boundaries have become increasingly blurred between industrial hazards within large industrial plants and industrial sites that are routinely imbedded in rural and suburban towns, fire departments are frequently supplementing their firefighting arsenals with extra weaponry. This petroleum-based fire occurred in the back of a truck storage yard surrounded by a commercial business, a railroad line, and a power utility substation. Several area fire departments acquired an airport crash truck as the core of a regional foam task force, which would be under their direct control, reducing a reliance on airports or military bases for help. Here, the foam response unit was put into service applying large quantities of foam to the storage yard and containers to prevent flare-ups or reignition.

Left: Foam application is always messy. Here, firefighters use a hand line to finish the foam application after the roof turret of the foam response unit was shut down. The roof turret provides a good overall blanket of foam, but firefighters with a hand line need to follow up to make sure that everything has been adequately coated with foam.

Industrial hazards confront municipal and volunteer fire departments every day. Many of the incidents involve trucks or trains. Here, members of the Stamford, Connecticut, fire department train with a propane prop to hone their skills with this type of incident. Training officers have the ability to light the fire under an old tanker to simulate a real-life event.

Opposite: Industrial-type fires can occur anywhere, so firefighters must constantly train to handle all types of incidents. Here, three firefighters advance an attack line on a propane tank fire as a means of learning how to handle an encounter of this type when it happens for real.

employees have predetermined locations to report to for their accountability. Radiation is not the worst fear of the fire brigade in many instances, due to their thorough knowledge of the plant and systems. If a release of radiation is verified, non-essential plant personnel and local residents are evacuated, while the essential workers are deployed from the safe areas to manage the necessary repairs. The fire brigade members respond to a radiation-safe area and await the proper time to enter the affected area, in the event that they are instructed to mend the problem.

RESEARCH FACILITIES

Research facilities have some of the same concerns and dangers as refineries and nuclear power plants, but they also contain different risks for firefighters. There are government facilities operated by the Department of Energy with conventional industrial hazards, hazards from experimental practices, and nuclear hazards. There are also the most frightening of hazards—those which are unknown because the nature of the materials being handled is classified. As odd as it may seem to the general public, secretive or classified research can impact the ability of firefighters

to handle or attack certain fires at these installations. An example would be a government research facility where knowledge of the nature of work being performed requires a security clearance that is not given to the firefighters. It is hoped that no emergencies occur in such facilities, but that does not eliminate the potential for disaster. Often, information about the substances that are being studied has not been made available to those outside the lab, and they are of an experimental nature, so firefighters do not know the level of danger or how to attack a fire within or around these labs. In some instances, the researchers themselves may not know how to extinguish a fire. In the case of a fire then, as rare and infrequent as they are, the firefighters must resist the sense of immediacy for extinguishing the fire and approach the situation in a methodical manner by communicating with the engineers or scientists to determine the appropriate course of action, which may include inaction. These facilities will have sprinklers and other automatic fire-suppression systems within the labs to attack the fires initially and hopefully mitigate the dangers before they get out of hand. In an extreme case, there would be no external attempt to fight the fire.

CHAPTER FIVE

WILDLAND FIREFIGHTING

Wildland fires used to be called forest fires. They may also be referred to as grass fires, brush fires, field fires, or vegetation fires. These fires often are unnoticed during their incipient stage, but a variety of factors can cause them to spread quickly to flammable vegetation and threaten structures.

What Are Wildland Fires?

The term *wildland fire* encompasses a variety of fires. There are surface fires that burn along the forest floor, ground fires that burn on or below the forest floor, and crown fires that jump along the tops of the trees, rapidly spread by high winds. Although some wildland fires are started by forces of nature, such as a lighting strike, most can be traced to intentional or unintentional human interface. Arson plays a large role in wildland fires. Some other common causes include an unattended or improperly cared for campfire, the careless disposal of smoking materials, heat from a vehicle's exhaust system that has been driven into dry grass, sparks from trains or electric transformers, or downed power lines. There are also spontaneous sources for wildland fires. Hay, manure, wood chips, and grain dust are examples of materials that can self-heat to temperatures sufficient for ignition. This process is accelerated with large piles or concentrations of these materials in areas where the air flow is restricted, and in high heat where the core temperature of these materials rises.

Wildland fires are influenced by the fuel that is involved and the topography of the landscape, as well as weather conditions including temperature, humidity, and wind. Each can factor greatly into the progression and spread of the fire, the suppression or containment tactics, and the duration of the event.

As mentioned earlier, some fuels burn along the ground. Depending on the weather conditions and types of fuel, the horizontal movement of the fire will spread to trees, which then allow the fire to begin spreading vertically. The fire will grow in both size and intensity by consuming fuels at all levels. This increases the radiant heat, which will quickly ignite surrounding flammable fuels, allowing the fire to grow rapidly.

A sample of the fuels located along the ground in wildland areas includes leaves, grass, dead wood, brush, tree roots, dry pine needles, and peat soils. Some of these fuels—such as small pieces of dead wood, pine needles, and dead leaves—allow for an especially aggressive fire spread, since they are loosely arranged, allowing for a free flow of air around them. Since these fuels catch fire easily, they act as kindling to ignite larger and heavier logs and trees. Some ground fuels retain moisture, however, and can actually slow a fire, since they are harder to ignite. As the ground fire grows in intensity, the flame heights increase and start the vertical extension into the taller trees. Not all trees burn the same. Pine trees, for example, burn faster than harder woods, such as oak. As these fires grow in intensity, air movement rising with the flames (known as convection heat transfer) can combine with the prevailing winds to create more wind, which increases the fire's spread further and faster. These winds can develop into a firestorm, which can create effects similar to a tornado or can develop into a microclimate (an atmospheric zone that is wildly different from the climate around it).

Normal winds can, of course, spread the flames through the trees or brush. They can also complicate

Flames burn freely through heavy timber in a forest area during the 162,000-acre "Day Fire" that burned in both Los Angeles and Ventura counties in September 2006.
Keith D. Cullom

the efforts of firefighters by carrying stray embers to unaffected areas, igniting additional fires. The wind also acts as a means to determine the physical direction of the fire, which is especially significant in areas with urban interface. As an example, the threat to homes and lives is diminished if the winds are blowing away from homes and into the forest lands. Winds blowing toward developed areas may create the need for mass evacuations and additional resources for structure protection.

An area with hills and canyons has different considerations than flatlands. As a fire enters a canyon, the upward burning of the flames and the intense heat quickly ignite the new fuel in its path, and the

fire spreads rapidly uphill. Another factor to be considered in hill fires is that the hillside that normally has direct exposure to the sun will be drier and therefore more combustible than the side that experiences more shade. High temperatures and low humidity are a bad combination for fire conditions. The two phenomena have similar effects on fuels. They both cause the moisture within the fuels to evaporate, making them drier and more susceptible to ignition.

WILDLAND FIRE RESOURCES

Wildland firefighting is unlike structural firefighting in many ways. First and foremost, the basic philosophy behind wildland firefighting involves perimeter control. Eventually, someone on the ground has to circle the entire event to either extinguish the fire or to ensure that it does not spread. The larger the size of the area involved, the more difficult and time-consuming the process will be. Once a fire has been surrounded by some type of fire line or fire break, it has been contained. At this point the fire is still burning and might still have the potential to spread beyond a fire line. The fire is controlled when the threat of spreading has been eliminated.

The largest fires, which receive the most media coverage and become the most well known, do not allow for 100 percent mop up, known in structural firefighting as overhaul. The area involved is just too large, and the firefighters cannot haul enough water for every spot fire. Hand crews spend weeks on the ground covering the burned-out areas using shovels and rakes in an effort to ensure that the resulting hot spots do not spread.

During the peak seasons with the greatest risk of wildland fires, the fire service works very closely with the National Weather Service to continually examine the variables that weather represents in the risk equation. Temperature, humidity, and wind are the most important factors to monitor. As important a role as the weather represents, the different regions, parks, forests, and areas with urban interface each present different factors to consider in terms of the fuel that will burn. Grasses, trees, and brush have different burning characteristics and, as a result, require different means from which to attack the fires that consume them.

Los Angeles City firefighters attack fire burning in heavy fuels with a 1½-inch hand line as flames race up-slope behind them. Equipment carried on their web gear includes a fire shelter, canteen, and other tools used for fighting a fire in brush or forest areas. *Keith D. Cullom*

Bulldozers are part of any first-alarm wildland response in California. The California Department of Forestry and Fire Protection has a statewide fleet of 62 bulldozers. They are used to cut a fire line to act as a boundary to contain the fire. The bulldozer blades the ground down to the bare soil. Specific fire conditions dictate if a 'dozer line is one 'dozer wide or needs to be the width of multiple 'dozers. One person staffs the transport and runs the 'dozer.

An engine pulling up to a small roadside grass fire might hit the flames from one spot with a hand line, or the firefighters might apply water while driving around the fire if the engine is equipped with pump-and-roll capabilities and a remote-controlled turret. The firefighters might or might not need to go into the grass with shovels or other hand tools.

A large fire will involve the use of bulldozers, air tankers, helicopters, hand crews, hotshot smoke jumpers, and hundreds, if not thousands, of firefighters. These assets will be discussed later in this chapter. If the fire is too big or too hot, then firefighters must concentrate on indirect tactics to control the perimeter, since they cannot directly fight the fire.

On a national level, several agencies are actively involved in the fight against wildland fires. These include the U.S. Forest Service, the Bureau of Land Management, the Bureau of Indian Affairs, and the National Park Service. Many states, as well as local jurisdictions that are prone to wildland fires, also maintain capabilities to control wildland fires. Wildland fires occur in all sizes. Significant fires can take place in Arizona, Idaho, Montana, Florida, Oregon, and California. Although enormous fires break out throughout these states with vast natural reserves, by and large the fires in California are a different breed. The reason for this is that no other state has the magnitude of urban interface into the wildland areas. Combine this with the enormous size of the state, and you have conditions that present some serious challenges to firefighters.

Therefore, it stands to reason that on a state level, the largest firefighting agency is in California. It is the California Department of Forestry and Fire Protection (formally known as CDF but now called CAL FIRE). Although CAL FIRE is best known for wildland firefighting, the agency also serves the residents of California in a capacity that encompasses all aspects of fire protection and rescue, and as EMS providers through local contracts in areas without their own municipal or county fire agencies.

Wildland fire responses consist of traditional engines, as well as engines designed more specifically for the wildland duties, with off-road and pump-and-roll capabilities. Pump-and-roll engines are specialized vehicles that have the ability to actively pump water while moving. Another crucial vehicle is the bulldozer, which is used for clearing brush or other fuels, creating a buffer that breaks the continuity of the fuel. This is known as a fire break, a fire line, or a fuel break.

continued on page 121

During a brush fire in the Topanga Canyon, a Los Angeles City Fire Department helicopter (Fire 3) makes a water drop on a spot fire racing up a canyon toward multi-million-dollar homes in the Bell Canyon area of Ventura County, just west of the Los Angeles County border. The two homes that were threatened by the fire were saved thanks to the quick and accurate water drop by the 'copter crew. *Rick McClure, LAFD*

Chapter Five

Right: Fires do not occur only in drought conditions. A bulldozer cuts a fire break through dense brush in central Florida. A slight haze can be seen in the background highlighting a fire in the dense woods.

Below: HME built several pumpers and an aerial for the Prescott Fire Department in Arizona. This Silver Fox unit is equipped with a remote-controlled front turret and pump-and-roll capabilities. Firefighters can drive and discharge water at the same time to surround a small vegetation fire. The officer has a joystick controller to direct the monitor where needed.

continued from page 118

Air support comes in the form of helicopters and fixed-wing air tankers. (Areas of the country that do not have large wildland fires often refer to trucks that carry large amounts of water as tankers, whereas those rigs are instead called tenders in regions that know a tanker to be an airplane.) The U.S. Forest Service maintains a fleet of more than 40 aircraft and has more than 800 aircraft available via contracts for use in firefighting and support operations.

Helicopters are used as a means to bring large amounts of water or fire retardant to the fire either with a fixed tank underneath the fuselage or with a large bucket that is suspended underneath. Helicopters are also used for aerial reconnaissance (often providing infrared imaging), to ferry helitack crews into the involved area as an initial ground crew, and as an emergency medivac for injured firefighters. The helitack crews, usually a crew of 10 firefighters, can rappel to reach fires in remote locations.

Firefighters can be deployed from engines, hotshot crews, helitacks, ground crews, and smokejumpers. Smokejumpers are specially trained firefighters who can reach fires in remote areas by parachuting in. This provides a rapid deployment of self-sufficient firefighters in rugged terrain who are ready to work without having to take the time or expend the energy to hike into the area. Tools, food, and water are dropped by parachute to the firefighters, making them self-sufficient for the first 48 hours of the firefight. Smokejumpers work seasonally, generally between June and October. They are based in California, Idaho, Montana, Washington, Oregon, and Alaska.

Hand crews from CAL FIRE are made up of California Department of Corrections (CDC) inmates, while other agencies may have career firefighters and seasonal firefighters. A crew consists of 18 to 20 firefighters. Their primary responsibility is to construct a fire line around the wildfire. This is a strip of land that they clear of flammable materials

A fast-moving brush fire crests the mountaintop behind homes in the Santa Clarita area of Los Angeles County. Dense smoke caused difficult breathing for many of the residents fleeing the neighborhood. Fire companies from the Los Angeles County and City fire departments moved into the area, provided structure protection, and were able to save the homes in this housing tract.
Rick McClure, LAFD

A U.S. Forest Service contract helicopter fills its belly tank from a convenient water source while working on the 58,000-acre "Ranch Fire" that burned in October 2007 during a Santa Ana Wind event in Los Angeles County. The two-engine Bell B-212HP helicopter was built in 1973 and fills its fixed belly tank through an attached suction hose.
Keith D. Cullom

down to the mineral soil. The fire line is intended to rob the fire of fuel so that it can proceed no farther. The hand crews are also tasked with the mop-up when the fire is out. The members of a hand crew have different tasks and specialties, which include the vital services of a lookout that monitors both the progress of their work and the progression of the fire. They use chainsaws, shovels, and other hand tools to create the fire line. They may travel long distances on foot and generally work a 12-hour shift. Hand crews are self-sufficient and may or may not have a base camp to return to. Often they make camp at their location at the end of a day. These crews are assisted by tankers and helicopters to slow the fire's progress so that they can stay ahead of the fire. They spend time scratching out a line or removing debris from the fire's path. During mop-up operations, they might spread soil that was moistened by water drops from aircraft to smother burning or smoldering areas that were missed.

Depending on the size of the fire and the type of fuel in its path, a fire line could range from a width of a few feet to an area of more than 100 feet. If the wind and other conditions permit, these larger fire lines are often created with backfires. These are fires intentionally set by the hand crews in an effort to rob the approaching fire of the fuel in its path. If the combustible material is depleted, the fire should die. These firefighters can be in severe danger if conditions shift, and it is always important for them to have a safety zone or area to get out of the path of the approaching fire. The hotshot crews are also hand crews and consist of the most experienced, multi-skilled firefighters.

Engine crews can vary from a minimum of three firefighters to as many as six firefighters. They may be assigned tasks similar to the hand crews, structural firefighting operations with hand lines for the protection of structures threatened by the wildland fire, or the suppression of structure fires as a result of the wildland fire. When a hand crew scratches out a line to light a backfire, engine companies may act to support the operation to ensure that the fire stays within the perimeter established by the hand crews.

Wildland Firefighting

Left: Firefighters in Prescott, Arizona, use shovels and fire rakes to demonstrate how they clear flammable vegetation to create a fire break. Their shirts and pants are made of Nomex for protection from fire, and they carry backpacks with supplies to sustain them during a prolonged period of work. The road flares would be used to create a backfire. The pouch labeled FSS contains a personal fire safety shelter. This is a small tent-like enclosure to be used by firefighters caught by a rapidly advancing fire. Usage of the FSS is a last resort for firefighters who cannot escape.

Below: The crew of San Diego City Brush Engine 40 rushes to grab tools and hose lines to attack flames roaring out of a drainage ditch during the 2006 "Day Fire" that burned more than 162,000 acres in both Los Angeles and Ventura counties. The 4 fighters assigned to Brush 40 were among the more than 4,500 firefighters who battled the fire for more than a month. *Keith D. Cullom*

Above: CAL FIRE is a state fire agency with more than 336 engines. Engine 2390, shown here, is stationed at the Washington Ridge Conservation Camp, 10 miles east of Nevada City. This is an inmate camp. All CAL FIRE hand crews are made up of inmates. The rig is mainly for camp protection but is moved during the summer to act as a reserve unit in nearby stations. This is an older unit, built in 1979 by Paeolitti, and is designated as a Model 9, which has a 500-gpm pump and carries 650 gallons of water. This particular engine has an International 1950 chassis, but the Model 9s also were ordered on Mack and Ford chassis.

Right: Pierce built three of these Model 25 urban interface units for CAL FIRE. The Pierce model is a Hawk. These have foam systems and the International 4400 Crew Cab. PPE for a wildland fire differs from structural PPE. The wildland Nomex suits are much lighter weight than structural bunker gear, and wildland firefighters do not wear SCBA. The firefighters in this photo are displaying an assortment of the wildland tools.

CAL FIRE provides fire protection by contract for areas that are not under the jurisdiction of the state. This 3,000-gallon tender serves the South Santa Clara County Fire District under contract by CAL FIRE. The tender carries water to support firefighting operations in areas that do not have traditional fire hydrants. The tender can support wildland operations by filling helicopters or supplying engines.

CAL FIRE Resources

CAL FIRE, as well as fire departments throughout California and surrounding states, models its wildland interface engines after designations set forth by the Incident Command System (ICS). The ICS designates three types of engines, and within CAL FIRE there are over 20 model descriptions to indicate design differences. The typing focuses on the overall firefighting capabilities of the unit, pump capacity, hose complement, and number of personnel. A Type 1 engine is a traditional structural rig with a minimum pump capacity of 1,000 gallons per minute (gpm) and normally is designed to stay on the road. It has a minimum of 400 gallons of water and a minimum crew of four. A Type 2 rig has less pumping capacity (500 gpm) and may have the ability to operate off-road. Like the Type 1, this carries a minimum of 400 gallons of water but has a smaller minimum crew of three. A Type 3 requires even less pumping capacity (120 gpm), although today most Type 3s have 500-gpm pumps. These routinely have off-road capability, are required to carry at least 300 gallons of water, and have a minimum crew of three.

The CAL FIRE model designations currently span several generations, each having specific characteristics or design changes from its predecessor. Type 3 models, for instance, included the Model 1 and the Model 5, both designed in the 1950s. These models were the backbone of the department for many years, with 500 gallons of water and a two-door cab. Crew seating was in the rear of the body on open-air benches. Many of these were purchased, most using the Navistar commercial chassis with bodies built by West-Mark, Master Body Works, and Westates. In the early 1970s, the Model 9 was designed with a crew cab, larger water tank, and urban or structural considerations, since the CDF was entering into contracts with local towns to offer services beyond wildland fires. More than 50 of these were put into service. They were purchased with Mack, Ford, or Navistar chassis with bodies by West-Mark, B&Z Truck Bodies, or Paeolitti. These units had larger, 650-gallon water tanks. In the mid-1970s, the CDF designed the Model 11. It carried more water and fewer men than the other units. The Model 11 came

Brush 7 in Corona, California, is a Type 3 urban interface unit built by E-ONE at their Superior plant in Canada. The chassis is a 4x4 International model 4800. The unit carries 500 gallons of water, 25 gallons of foam, and a 500-gpm rear pump. One benefit of the rear-pump design comes into play when operating at roadside brush fires. This positions the pump operator safely out of traffic regardless of which shoulder the rig is parked on. Additionally, all of the preconnects are off the rear, with the exception of the red line, which is on top and can be deployed to either side without endangering firefighters.

in with a Mack, Ford, or Navistar chassis with bodies by Master Body Works, West-Mark, or Paeolitti. Roughly 20 of these entered into service.

In 1990, the Model 14 and Model 15 engines were introduced with four-door crew cabs, bringing all the firefighters inside the cab. These units differed only in that the Model 14 offered two-wheel drive and the Model 15 had four-wheel drive. They were both purchased with a four-door Navistar chassis and bodies provided by Master Body Works, West-Mark, or Placer Fire Equipment. Both models had 500 gallons of water and 500-gpm pumps. These were to become the fleet staple for many years, with 107 two-wheel-drive and 81 four-wheel-drive units going into service.

The Models 16, 17, and 18 all introduced custom fire truck cabs and were all geared more toward the urban interface than wildland firefighting. None of these units were bought in large quantities. Models 16 and 17 had Spartan chassis with bodies by Westates. Four Model 16s went to the fire training academy and four Model 17s went into service. The Model 18, a Type 2, had an HME chassis, a West-Mark body, 600 gallons of water, and a 1,000-gpm

Above: Firefighters prepare to make a stand in an effort to hold flames from jumping Placerita Canyon Road during the 6,000-acre "Foothill Fire" that burned in Santa Clarita in July 2004. More than 1,400 firefighters from 200 engines worked to control the fire. *Keith D. Cullom*

Left: This rather unique-looking rig is from Laredo, Texas. Built on an all-wheel Freightliner M2 chassis, it is a Brushmaster model off-road vehicle by Rescue Master, which is a division of American LaFrance. Laredo is on the U.S. border with Mexico and has a large amount of rough wilderness terrain within its district. The city has three of these brush units in service with pump-and-roll capabilities, under-truck ground nozzles, a remote-controlled bumper turret, dual red lines, and a full cab brush guard to protect the metal and glass from trees and heavy brush. The majority of the body was designed with diamond plate instead of painted metal since the unit takes considerable abuse. The diamond plate stands up better over time than a painted surface to the scratching from branches and shrubs. These units carry 1,000 gallons of water and have a 350-gpm pump.

tank. The Model 19 was similar to the previous custom rigs, which were either built on HME or Spartan chassis. The body was by Master Body Works. This unit had the larger 600-gallon water tank and a 1,000-gpm pump along with more ladders, increasing its functionality for urban interface while still providing off-road use for wildland incidents. Six of the Model 18s were purchased, and only two of the Model 19 units were purchased.

In 2001, CDF experimented with two prototype rigs from Pierce, going back to the four-door Navistar four-wheel-drive chassis. These had 500 gallons of water with 500-gpm pumps and were field-tested. No purchases beyond the original two were made. The Model 25, also a Type 2, was a similar unit from Pierce with a 1,000-gpm pump. Three of the Model 25s went into service. The most current CAL FIRE Type 3 engines are the Models 34 and 35. Like Models 14 and 15 of the early 1990s, these new units differ only in that the Model 34 has four-wheel drive and the Model 35 has two-wheel drive. They are on the newer Navistar four-door chassis with body work by Placer Fire Equipment. CAL FIRE has gone back to 500-gallon water tanks and 500-gpm pumps.

All CAL FIRE apparatus are required to carry enough supplies and equipment, including food and fuel, so the crew can be self sufficient for 24 hours.

FIGHTING WILDLAND FIRES

Perhaps the most recognizable dangers in the area of wildland fires are the warm and dry Santa Ana winds, which arrive during autumn and early winter in southern California, and the north winds in northern California. The fact is, though, that many items combine to create conditions that cause the state and federal agencies to issue high-danger warnings for the risk of wildland fires.

A typical response in California to a report of smoke showing, whether it is reported through a 9-1-1 call or a fire lookout, is based on the weather conditions. These and the status of the different fuel types in the area are constantly monitored. A first response could involve five engines, two bulldozers, two or three tankers, one air attack aircraft, one helitack, two hand crews, and one chief officer. The response can be altered based on the historical needs to control fires in the particular area within the first burning period. The first burning period is defined as an area of 10 acres or less that can be extinguished that same day. Most fires are contained in the first burning period.

The officer on the first-arriving engine becomes the IC and has to develop a beginning plan. The IC is responsible for directing other resources to accomplish the mission at hand and for understanding and interpreting the behavior of the fire.

Firefighting strike teams arrive on Pacific Coast Highway at the scene of the "Corral Fire." They stage as they wait to receive line assignments for protecting structures in the coastal canyons of Malibu that were threatened by the fire that burned nearly 5,000 acres and destroyed 53 houses. An engine strike team is made up of 5 like engines and a leader. More than 300 engines were used to contain the fire. *Keith D. Cullom*

The task of containing wildland fires in the first burning period is complicated in areas where homes have encroached on the wildland areas. The IC and first-arriving units must commit themselves to investigating the fire's effect on homes and residents, in addition to trying to accomplish the important task of perimeter control. If it is possible to abandon the structure protection, it might be possible to contain the fire and reduce the threat to structures. Resources that are directed to protect structures and draw down what is available for perimeter control may allow the fire to gain in strength. If the houses become involved, an increased number of resources are required to protect the exposures, which can become overwhelming to manage until a suitable increase in command personnel can be deployed. Studies have shown that a mind can handle close to a dozen things at a time under normal conditions, but under stress the number decreases to three or less. The need for additional command support in this situation, therefore, becomes evident.

Once the initial attack assignment has been put to work, the IC is tasked with requesting additional resources by type and numbers. In the event of a large-scale fire, it is common for an IC to place a single request for hundreds of personnel and vehicles to be assembled. These might come from across the state or even neighboring states. To simplify the deployment of these resources, the agencies have predetermined strike teams, which are requested as an initial need, on an immediate-need basis, or as part of a planned need. The initial attack strike team requests require units to get on the road immediately and assemble with the rest of the strike team close to the incident. The immediate need requires the teams to be on the road and en route within 30 minutes of the request. These units can expect to be placed on the fire line immediately upon arriving at the fire. The planned-need request is intended to cover a future need at a specific place and time. The planned-need units are relief companies that will be deployed at a predetermined time to replenish the crews that have been working constantly prior to that time.

A strike team consists of five resources of the same type and includes a strike team leader. The strike team leader takes directions from the IC and manages the resources, including the responsibilities for food, supplies, safety, and welfare of the crews. During the fire season, agencies have pre-assigned personnel and equipment to respond with a strike team if they are called upon. Individuals are responsible to have with them enough personal gear to survive for three to seven days. The mutual aid agreements that govern

A CAL FIRE Grumman S-2T Air Tanker delivers its 1,200-gallon retardant drop onto an open flank of a wildland fire in Santa Barbara County, California. The state agency operates 23 S-2s in its fleet of tankers, which are assigned to 13 different air attack bases throughout the state. These were purchased in 1996 by CAL FIRE (then known as CDF) from the Navy, who used them between 1958 and 1975. *Keith D. Cullom*

the response to these large incidents ensure that an agency is only to participate in the deployment if doing so will not adversely impact the primary mission of that agency, which is to provide fire protection to their residents. The cost in dollars for these responses is high. Depending on the distance traveled for the deployment, the engine type, and the salaries of the firefighters, a single strike team can cost $25,000 per day, or $5,000 per vehicle per day.

As a fire grows in magnitude, the requirements of the fire service increase exponentially. If an event will last more than a day or so, then the management

Chapter Five

Right: Large-scale incidents require an enormous commitment of resources and coordination between multiple agencies. Many departments have for years deployed mobile command posts to serve as a forward base of operations for the IC and support personnel. Since 9/11, federal money has become available for the purchase of sophisticated command-and-communications units capable of becoming self-sustained units providing space for meetings, joint communications, and the ability to connect multiple agencies or departments on different radio systems. This unit, which is part of the Arlington County Fire Department in Virginia, is one such vehicle. Each side of the body has a slide-out module that creates additional interior space. The extension visible on the driver's side is part of a meeting area with bench seating, a large conference table, flat-screen monitors, phones, and radios to allow the IC and other agency heads to work in an area that is separate and, if need be, private

from the multiple radio operators in the forward section. Arlington County added dual exterior message boards for the ability to convey messages to the public or the media that might be in the vicinity of the command post. Stabilizer jacks underneath are needed when the slide-out modules are deployed. Exterior cameras, a weather station, and satellite receivers are located on the vehicle's roof.

Below: A view of the forward section of the Arlington County command and communications vehicle shows six work stations with phones, radio consoles, computer terminals, and monitors, which can be programmed to display satellite feeds, camera feeds, or computer information that is relevant to the incident at hand. The three gray seats on the far right are affixed to the slide-out section, which illustrates quite clearly the added space gained from this design. When the section is retracted, the visible aisle disappears. The consoles closest represent multiple monitors to view satellite images from various news outlets, camera images from the multiple cameras around the vehicle, or other feeds from cameras deployed at an incident. The controller box on the counter allows the operator to position the mast and camera, which telescopes high above the roof. Additionally, all signals coming into the vehicle can be recorded for future review.

requirements expand to include food, shelter, relief personnel and supplies, equipment maintenance, fuel, on-site information—including weather and condition updates, as well as status of the fire—and medical needs. The incident managers have at their disposal the resources to provide catering units, tents, and mobile offices to support the incident action plan governing the event. The mobile offices are necessary to handle the mountains of paperwork required to manage a large-scale fire. Initially, CAL FIRE has resources for kitchens that are staffed by CDC inmates. When these resources are depleted, private suppliers are available to handle the catering and offices through national contracts with the federal government. If hundreds of vehicles are called to the scene, there will be mechanical breakdowns and flat tires, which necessitate a working area for mechanics and their tools. Portable weather observers from the National Weather Service are available to assist on-scene to pinpoint the weather for the incident.

Communications for a major incident are crucial for the safety of personnel and the successful coordination of resources to mitigate the fire. The rough terrain can cause problems with radio traffic, which dictates the need for portable, programmable repeaters to keep multiple lines of communication open.

Left: The Chino Valley Fire District in Arizona has three styles of Silver Fox pumpers from HME. This unit has a 1,500-gpm pump and carries 750 gallons of water. The high-set cab is on a 4x4 chassis to allow off-road access in a largely rural district. Wildland firefighting is an integral part of training, due to the vast natural lands and open areas in this dry climate. Here, with a summer monsoon approaching in the distance, firefighters use tank water to wet down an area off the highway. Small off-road fires within 100 feet of a drivable area are handled with the preconnected front bumper line and are quick work with the ability to maneuver the Type 1 engine into shallow brush.

Left: Some of the more than 220 engines assigned to the "Pine Fire" in Los Angeles County stage at the fire base camp during their off shift. Typically, firefighters assigned on the fire line work 12- or 24-hour shifts on the line and then are assigned a rest period at the base camp. The Pine Fire burned 17,500 acres in July 2004. *Keith D. Cullom*

Evacuations may become necessary as an event grows in magnitude. Evacuation requires the notification of the public and the mobilization of law enforcement personnel, which may even need to be supplemented by the military. The media is one means to notify the public of the need to evacuate. The emergency dispatch centers have the ability to utilize a reverse 9-1-1 system, which allows them to send thousands of warning messages simultaneously to land lines and cell phones based on the 9-1-1 system. These enhanced 9-1-1 systems are able to track incoming calls and are able to reverse this capability to send messages along the same routes. Although incorporation of landlines is a natural part of 9-1-1 systems, cell phone participation for the reverse 9-1-1 system is voluntary and requires registration on the part of the consumers.

Above: This surplus military 6x6 serves the Yaphank Fire Department on Long Island to handle fires in the off-road brush and trees. The rig has a tank and a monitor. It goes right into the woods regardless of whether there is a road or not so firefighters can get right at the fire. The brush guards and large tires allow it to be driven through almost anything it can find. Once the tank is empty, it must be refilled by a tender.

Consider this historical look at what is known as the 2003 Southern California Fire Storm as an example of the magnitude of resources that the state of California is able to assemble and manage in the event of large-scale fires. Between October 21 and October 30, 11 separate fires were demanding resources. These fires affected San Diego County, Riverside County, San Bernardino County, Ventura County, and Los Angeles County. In total, 3,641 homes were destroyed, and the full combined area involved 736,860 acres. There were 214 injuries and 22 fatalities. The deployment involved 14,000 personnel. The largest single fire in this incident was known as the Cedar Fire, which was one of four in San Diego County. This fire accounted for 113 injuries, 14 deaths, the loss of 2,232 homes, and the involvement of 273,246 acres of land. Although the management of this event was not without its flaws, it illustrates the ability of the state to muster and manage an incredible amount of resources.

Right: Los Angeles City firefighters advance a 1½-inch attack line in an effort to knock down a wildland fire burning in heavy fuels in Griffith Park. Equipment carried on their web gear includes a fire shelter, canteen, and other tools used for fighting a brush fire or a fire in forest areas. *Keith D. Cullom*

CHAPTER SIX

RESCUE WORK

Firefighters do much more than fight fires. They are also tasked with providing various types of rescue services. In fact, many fire departments have rescue as part of their name. It is not uncommon to see "Fire and Rescue Department" on rigs, patches, letterhead, and fire stations.

The rescue aspect of a firefighter's job can encompass many types of events, ranging from minor incidents like freeing a child entangled in a banister to rescuing someone from a body of water, saving someone trapped in a collapsed building, or extricating someone from an auto wreck. High-angle rescue deals with saving a victim high above the ground as opposed to below-grade rescue, which occurs below the ground. Other types of technical rescue include what is referred to as urban search and rescue in the aftermath of building collapses and other damage incurred as a result of storms, earthquakes, fires, explosions, and other phenomena that result in victims (alive or dead) being trapped by various debris. There are highly technical aspects involved with training for each type of rescue and utilizing the resources available to firefighters. Often, rescue work is divided among teams with special training in the various tasks. In addition to the rescues listed above, there are also water-based rescues that include diving, ice rescues, and swift-water rescues.

RIGS

The tools and equipment needed for these special teams are generally carried on dedicated rigs or trailers. Although first-responding engines, trucks, and rescue squads often carry small amounts of equipment to initialize rescue work, the substantive resources will arrive with the team members. Depending on the type of rescue equipment or the type of company, the rigs will be tailored to their needs. Rigs for the special teams are often not staffed unless there is an event that requires them to respond.

The most common of these rigs is referred to as a rescue, squad, or heavy-duty rescue (HDR) unit. Different departments use any one of these terms to identify a rig that traditionally carries no water, hose, or ladders. It is ideally staffed with a large company of firefighters who can perform specialized rescues or supplement manpower at the scene of a fire. The HDR is often a large rig with massive amounts of storage space to accommodate all of the tools and equipment necessary to handle most emergencies that the fire department will be called upon to handle.

This equipment may include, but is by no means limited to, some or all of the following items: hydraulic rescue equipment, saws, torches, hand tools, generators, cribbing, air bags, a thermal imaging camera, various detection meters, supplies and suits for handling hazardous materials, dive gear, spare air bottles, rope, rigging, and other cutting tools. Many departments devote special attention to the design of these rigs to ensure the most efficient use of the available space by tailoring the specific mounting of every tool and piece of equipment. Over the past several years, many of these rigs have grown significantly in size and complexity. The cost of the vehicle itself can represent only a portion of the total expense when the value of the tools and equipment is added. Some of these rigs are so large that they have a separate tractor and trailer to improve maneuverability. The complexities involved with

Opposite: Three rescue technicians use a diamond-tipped chainsaw, hammer drill, and chipping hammer to breach an 8-inch reinforced bridge deck. A variety of tools are used simultaneously to facilitate a quick rescue. While some tools work more efficiently than others, each has its own unique attributes and, when combined, all complement each other.

The Phoenix Fire
Department received three
of these special operations
squads from American
LaFrance in 2006. These
were part of a larger order
for the state of Arizona,
which placed similar
units in Glendale, Tempe,
Chandler, and Tucson.
Purchased with state funds,
the individual cities staff
the units and are free to
use them as they wish, but
a condition for ownership
is that they are available
to assist with a large
deployment anywhere in the
state with personnel from
the department that owns
them.

urban search and rescue have increased the tools
and equipment carried by many fire departments.
Often, these added supplies require the addition of
supplemental vehicles and separate trailers.

VEHICLE AND MACHINERY RESCUE AND EXTRICATION

Auto accidents are the most common type of rescues
for firefighters, possibly even more so than rescuing
victims from fires. A patient has to be extricated from
a vehicle if access to the patient is blocked due to
damage from the crash. This could be as minor as
requiring a door to be popped open with a pry bar or
Halligan tool, or as significant as a need for multiple
hydraulic tools to cut the vehicle apart. While the
general public and the media refer to all hydraulic
spreaders generically as the "jaws of life," this is actually a brand-specific name that has come to symbolize
all similar tools, much like the word Kleenex is used
to refer to all types of tissues.

Hydraulic tools are powered either by a portable
generator that is carried close to the extrication work

being performed or by internal, on-board generators
built into the rigs. These generators supply power
via preconnected hose reels, which often allow
firefighters to work at distances of up to 100 feet from
the rig. Among the arsenal of hydraulic tools are the
spreaders, cutters, and rams. Each has its own uses
during the extrication. The spreader, which has two
arms that open to form a "V" shape, can be wedged
into a tight area and force objects apart, allowing
firefighters access to a patient. Cutters use hydraulic
power to cut steel like a giant scissors. Rams are
compact hydraulic tools that extend to force a
wider opening between two materials than can be
accomplished with spreaders. Small rams can be used
in areas too small for a bulkier spreader. Rams can be
utilized in spaces that range from 15 inches to more
than 2 feet. The largest rams are capable of creating
a space of almost 5 feet. By comparison, the largest
spreaders can create an opening of up to 40 inches.

Other tools that can be used in extrications
include reciprocating saws, power chisels, and a host
of conventional power tools, plus cutting torches, air

After the patients have been removed, the cleanup phase begins. Firefighters put away all of the tools and equipment that they used to perform the rescue. This firefighter is releasing a hydraulic ram that was used to widen the opening on the driver's side, which gave them more room to remove the driver. A hydraulic spreader and cutter can also be seen on the ground at the rear of the car.

As one team of firefighters cuts the hinges on the door of this overturned car, others are bringing more tools and equipment off the heavy rescue unit. Since the rig is fairly close to the rescue operation, the hydraulic cutter is connected to a line that is coming directly off the rig. A hydraulic spreader was used to create an opening for the cutter. Some departments use wood cribbing for stabilization of the car, but this department chose synthetic cribbing, which is made from a hard polymer.

Bellmawr, New Jersey, located in Camden County, is the home of this 2007 Seagrave heavy rescue. Company 32 is a volunteer department, and they designed this unit to put every bit of available space to good use. A partial listing of what they carry includes a full complement of hydraulic tools and cribbing on each side of the rig along with air bags, torches, air bottles, and saws.

bags, bottle jacks, air chisels, Porta-Powers, and chain come-a-longs. An important aspect of any extrication is the stabilization of the vehicle to prevent it from moving or shifting. This can be accomplished in a variety of ways using pneumatic shoring jacks, cribbing, chocks, and air bags.

As the extrication proceeds, medics need access to the patient to begin rendering aid and to remain with the victim to provide reassurance as the rescue progresses. If possible, the medics will be inside the vehicle alongside the patient, perhaps under the protection of a tarp to protect both from glass or other loose debris that might cause injury. Depending on the level of damage to the vehicle, the extent of the entrapment, and the degree of the patient's injuries, the firefighters might have a small amount of work or may have to take the vehicle almost totally apart for the proper access and working room to remove the patient. Extrications can be completed in a matter of minutes or can be extremely complex, resulting in lengthy or prolonged rescues.

URBAN SEARCH AND RESCUE

Urban search and rescue (US&R) encompasses locating, preliminarily stabilizing, and extricating victims trapped in building collapses. Earthquakes, hurricanes, floods, tornadoes, and terrorist attacks are major events that may require the assistance of US&R task forces. The addition of terrorist attacks to the Federal Emergency Management Agency (FEMA) deployment list was a result of the bombing at the World Trade Center in New York City in 1993 and the Murrah Federal Building in Oklahoma City in 1995. US&R teams may be organized on the federal level under FEMA, at the state level as part of a state's emergency management agency, or on the local level as a fire department US&R or technical rescue team. Most of these different teams operate under similar guidelines in terms of protocol and deployment.

There are 28 FEMA US&R task forces represented in Arizona, California, Colorado, Florida, Indiana, Maryland, Massachusetts, Missouri,

Nebraska, Nevada, New Mexico, New York, Ohio, Pennsylvania, Tennessee, Texas, Utah, Virginia, and Washington. Florida and Virginia each have two teams, while California has eight. FEMA teams are currently serving under the auspices of the Department of Homeland Security (DHS) and are deployed to large-scale national events. In some cases, specific teams may deploy internationally. Prior to the terrorist attacks of September 11, 2001, FEMA was an independent agency. After 9/11, when the administration felt that terrorism was a greater risk than natural disasters, it was rolled into the new DHS. This change brought additional bureaucratic layers placing the military over FEMA.

Components of these teams are always on standby and must be able to deploy within 6 hours and be self-sufficient for more than 72 hours once on-site. A FEMA task force has to have three sets of staff that are capable of filling 70 different positions. Varying by task force, they may be broken down into three teams that are on call for three months at a time. Each fire department member has EMT certification and belongs to one of six components of the team: search, rescue, hazardous materials (hazmat), medical, logistics, or planning.

In 2001, the fire department in Provo, Utah, purchased several new units that were built by Pierce on the Quantum custom chassis. This Encore model special operations unit has roll-up doors on both sides, plus additional storage space in the rooftop compartments. All of the units in Provo are staffed with paramedics to cut down the response time for the multitude of EMS calls. The cab of this unit has a slightly different design without a crew entry door on the driver's side. This accommodates a command area with a desk in the rear of the cab.

Above: A hydraulic cutter is used to cut through the rear corner post of this car. When all four corners have been cut, the firefighters will lift off the car's roof, allowing them access to the victim inside. Before they can cut the posts, they have to remove the front and rear windshields.

Above right: The top hinge on the passenger door has been popped, and now the firefighter is using a hydraulic spreader to pop the bottom hinge so they can completely remove this door. The wood step cribbing underneath, in conjunction with several similar pieces, is meant to keep the car from moving.

The search component consists of technical search specialists trained in the use of sophisticated electronic victim-locating equipment, as well as canine handlers and their specially trained search dogs. The rescue component consists of specially trained rescue technicians, as well as heavy equipment and rigging specialists. The rescue operation has to protect the victims who are trapped, as well as the rescuers. The rescue section is tasked with the assignment of reaching the trapped victims, along with working with the engineers to implement the necessary stabilization

of the areas to be searched. Hazmat technicians are equipped to evaluate the collapse scene for potential contamination or other hazards that may be unsafe for the rescuers.

The medical component includes emergency room physicians, paramedics, nurses, and other staff who have the capability of providing advanced life support for victims and task force members if needed. Logistics section members are responsible for the over $3 million of equipment that the task forces deploy with, procuring local resources required to maintain equipment and personnel, as well as establishing and maintaining a robust array of telecommunications and data systems for task force use. Planning personnel are responsible for coordinating and documenting the action plans that the task forces utilize when they engage in operations, as well as structural engineers who assist with triage, safety assessment, and planning responsibilities. The engineers are tasked with having a commanding knowledge of breaching, shoring, and moving structural elements at a

The impact from this collision created quite a bit of front-end damage to this full-sized, American-made car. Firefighters from several suppression companies and the heavy rescue in Orange County, Florida, are working to free the driver. One firefighter prepares a hydraulic ram for use in the extrication to roll the dashboard in the event that the hydraulic spreader is not enough. The spreader has been set after forcing a large gap between the dashboard and the floorboard, which is resting on the pavement.

scene while the search crew uses cameras, listening devices, and dogs to locate victims. The engineers have to assess the relative dangers of the site and prepare a plan to initiate a safe search for victims and their subsequent rescue.

Generally the teams work 12-hour shifts, and FEMA teams usually remain operational for 10 days. If the incident requires a longer presence, additional task forces will be deployed to relieve the initial teams. Operational guidelines were modified after the Hurricane Katrina disaster in 2006. The predeployment time is no longer considered part of the operational period, so deployments extend to approximately two weeks. It should be apparent from the previous descriptions that these teams are not solely composed of firefighters but include many highly trained private-sector individuals to provide a cohesive unit to support and perform the search and rescue operations. The federal government pays for all expenses and salaries incurred with a federal deployment.

In a situation where the scene involves multiple structures, or large multi-room/-floor structures, searchers and rescuers utilize a marking system on the building exteriors to indicate when the structure was searched and by what team, what hazards are present, and how many victims are trapped inside. These

This is what the front end and passenger compartment looked like after the victim had been removed from the car and the dashboard had been separated from the floorboard. Firefighters must be able to adapt the use of their tools to each situation that they encounter. The amount of compression in the engine compartment created by the collision complicated a traditional dash-roll maneuver.

Not all techniques involve highly specialized equipment. After completing a four-sided cut in a slab, technicians use simple mechanical advantage and brute strength from several members to lift out the 500-plus-pound section that will allow them to enter the void and remove trapped victims.

Above: Prior to any cutting, it is important to determine if a known or suspected victim may be in the path of a cut. Not knowing the answer to this question may result in serious injury, decapitation, or death to a victim who is struck by a tool. These technicians have made a small inspection hole and inserted an adjustable, flexible camera to ascertain the victim's position. The camera offers video and audio.

Above right: Holding a 90-pound jackhammer is no easy task. These rescuers have rigged an adjustable strap to support the tool. This makes the work easier and safer. Other technicians monitor the activity and tool operator to ensure a safe and accurate rescue.

Right: Cutting steel in tight areas presents many challenges. Practice in training prepares technicians for this very task. Limited workspace requires technique and finesse. Anytime cutting operations create sparks, the threat of fire from ignition is always a concern. Fire extinguishers and hose streams are kept nearby, should they be needed.

Firefighters and EMS personnel in Milwaukee, Wisconsin, work to free a victim of this multi-vehicle rollover accident. Working simultaneously, the medic has prepared intravenous solutions to administer to the victim as firefighters begin to raise the pickup, enabling them to extricate the victim from underneath. While some firefighters brought the tools and equipment right up to the scene, others determine just what they need to complete the rescue. They have inserted and inflated air bags to begin lifting the truck. They have hydraulic spreaders and a large quantity of cribbing material to stabilize the truck.

labels are arranged around a large painted X and can further denote the location of the victims, whether they are alive or dead, and whether they have been subsequently rescued or removed.

State US&R teams usually follow the FEMA design in terms of training, personnel, and equipment. The state assets, though, have more flexibility with less red tape and bureaucracy. They do not require a federal mandate for deployment, which also means that they are not funded by the federal government. It is a state's responsibility to maintain adequate funding in the budget to sustain its team or teams.

In the event of any emergency, local fire departments must respond first. State teams are added subsequent to the fire department response, which are then supplemented by a request for federal resources to complement the first responders as the situation requires. Some states fund regional teams that are part of large, full-time fire departments. These teams are required to respond immediately to disasters within a designated area.

Many fire departments have their own US&R or technical response teams or have mutual aid agreements and affiliations that provide a similar immediate response to an event. As the need for more

resources becomes evident, either due to the magnitude of the incident or to replenish exhausted firefighters, requests are initiated through additional mutual aid on the local, county, or state levels. Subsequent to the events of 9/11 and Hurricane Katrina, federal funding has been increased through grants to fire departments throughout the country in an effort to bolster the ranks and abilities of first responders, namely local fire departments. This money has provided for extensive training, preplanning, logistics, equipment, and vehicles to outfit the first line of defense. This has proven to be a boon for the various industries that supply the fire service with these items. Enormous and fully stocked specialty apparatus can be found in hundreds of fire stations. Equipment is in trailers of all sizes, as well as straight-frame and tractor-drawn trucks to sustain the local US&R or technical rescue teams through minor or larger events, or until the state or federal teams arrive. In many cases, members of the local or regional teams may also belong to a federal team. If a disaster strikes in a jurisdiction that has a federal task force, it is more than likely that DHS/FEMA will not deploy that specific task force to the affected location when there is a request for assistance. In these cases,

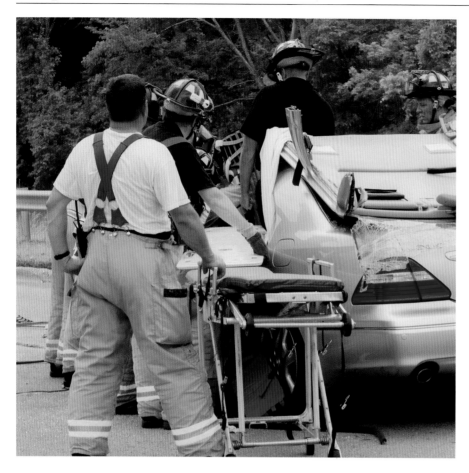

After removing the roof of the car to gain access to the patient, firefighters prepare to remove the car's driver for transport to the hospital. The patient has been strapped into a Kendrick Extrication Device (KED), which is like a backboard from the waist up that keeps the patient immobilized for removal to the stretcher. It is a soft, plastic brace with a rigid interior, that straps under the legs, around the waist, the chest, and the head. It is often used for a stable patient as an extra precaution to immobilize the head, neck, and spine to prevent further injuries.

that task force and personnel usually remain state assets under the direct authority of the state emergency management agency or other authority having jurisdiction. Regardless, although DHS/FEMA may predeploy assets in a standby mode prior to a known peril, such as a hurricane, the task forces cannot be mobilized to the disaster scene until a request is made by the governor of the affected state.

RESPONDING TO AN INCIDENT

At a major incident, the capabilities of the first arriving fire department personnel are often limited to the rescue of surface victims and the walking wounded. As the technical rescue teams arrive, they do initial structural assessments and begin to determine how many victims they might be dealing with, what condition the victims are in, and how that might change in the time it takes to reach them. The structural assessment deals with an evaluation of what needs to be done to create as safe an environment as possible to begin the task of search and rescue. FEMA has a matrix outlining the phases of a search and rescue at large events. The outline predicts that 50 percent of those people who are injured but not trapped will be rescued by civilian intervention—people help-

ing each other who are nearby at the time of the event. The next 30 percent—again victims who are not trapped—will be rescued by the first responders. Fifteen percent who are lightly trapped will be rescued by the local technical rescue teams, and the remaining 5 percent who are heavily trapped will be rescued or recovered by the US&R task forces.

Locating and reaching the heavily trapped is a slow and methodical process that requires search dogs, cameras, and vibration and acoustic listening devices. Once a victim's location has been determined, debris may be moved by hand and with buckets until the workers reach heavier debris. Then, an inspection hole may be made to allow the insertion of a camera to see the position of the victim so the rescuers know how to proceed. They may use saws, jackhammers, or hydraulic tools to move, cut, break, or breach the

debris. If the victim is safely away from the immediate area of entry, rescuers may utilize what is referred to as a dirty breach, which is more of a rough blasting or breaking up of the debris. In the event that the victim is closer to the point of entry, then a clean breach or lift-out will be necessary. Here, the workers cut the debris and lift it out of the way to prevent further injury to the victim. Certain types of building materials and methods, as well as certain failure/collapse systems, inherently create voids within or below the collapse. The presence of these voids lends the highest probability of locating living victims.

Technical rescue workers do not wear traditional firefighting PPE. They wear coveralls, jump suits, or military-type battle dress uniforms (BDUs) with strong boots and hard hats. Disaster sites contain many hazards, including falling debris and particu-

This Special Operations Heavy Rescue unit from Hillsborough County in Florida is the longest unit built to date by Saulsbury. Initially, the unit built in 2000 for the Packanack Lake Fire Company 5 in New Jersey was to be the longest, but this was surpassed by the Wayne Memorial First Aid Squad. Well, the Hillsborough County unit beat them both for the record. Aside from being long, it carries a tremendous amount of equipment and is staffed by firefighters who also belong to Florida Task Force 3, a Florida State US&R task force.

late dust, so protective eyewear is important, along with some type of mask or respirator. As with any emergency site, reliable radio communications is also of utmost importance.

WATER RESCUE

There are three basic types of water rescues for land-based firefighters. The first type involves traditional divers. Boating accidents, swimming incidents, people falling into bodies of water, and accidents

Above: Saulsbury built this heavy-duty rescue squad on an E-ONE Cyclone chassis for the Packanack Lake Fire Company in Wayne, New Jersey. This unit was the longest rescue built by Saulsbury when it was delivered. This claim to fame was short-lived since the Wayne Township Memorial First Aid Squad purchased a longer rig soon after this one was delivered, and both were topped later in the same year by a rig for Hillsborough County, Florida. Rescue 5 has an extended crew area to carry up to eight firefighters and has a command post in the body directly behind the cab. It is not uncommon for volunteer fire departments like this one to dress their rigs with hand-painted gold-leaf lettering and trim. The red lights on the inside of the cab allow firefighters to adjust to the lighting around them without having to acclimate their eyes from the bright harsh white light to the nighttime darkness when they exit the vehicle.

Below: Perhaps a once-in-a-lifetime photo opportunity, this shot depicts all five FDNY Rescue Company units that were delivered to replace the rescues that were destroyed on 9/11. Saulsbury built these rigs on the E-ONE Cyclone chassis with stubby cabs. The cabs only seat the chauffeur and the officer, while the rest of the company rides in the back of the rig where there is room for them to assemble tools and suit up appropriately en route to calls. Even though the five rigs appear identical, each rescue company customizes their own unit to meet their specific needs and requirements. Each rig also has unique decals illustrating the specific company logos. Rescue 1 on the far right can be seen with its inflatable boat stored on the roof.

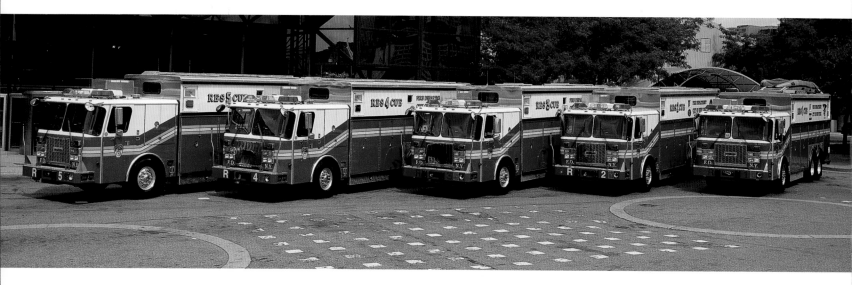

Left: Many fire departments have heavy rescue units or urban search and rescue (US&R) units, which carry an abundance of specialty equipment. These departments are capable of filling as much space as they have available and often need additional trucks or trailers to store all of the tools and equipment that they feel are necessary to have at their disposal. This tractor-trailer combination US&R unit is from Costa Mesa, California. The trailer was built by Hackney and designed to maximize space for the equipment. There is an equal amount of tools and supplies on the other side of the trailer, which also has a large generator, a telescoping light tower to illuminate nighttime scenes, and additional rooftop compartments. The vertical tool boards slide out for easy access.

where a vehicle becomes submerged in water are basic examples of the need for a response from a fire department dive team. On-duty divers may be able to suit up en route if they have a vehicle with the space to maneuver; otherwise they need to take the time on-site to get into their wetsuits, fins, breathing apparatus, and other gear. The divers always are supported by on-shore backup or tenders to monitor, tether, or keep track of the dive. Depending on the water conditions and temperatures, they may instead use dry suits, which provide thermal insulation against cold water.

Dive emergencies are classified as rescues or recoveries. Depending on the circumstances of the incident and the elapsed time, a rescue signifies the possibility of saving a life, while a recovery is the retrieval of a body or bodies. These rescues can involve boats, multiple divers walking in shallow water or being pulled along behind a boat with a tow bar, or the use of side-scan sonar to locate the victims. Currents can be unpredictable and make it difficult to locate someone in relation to the last position that they were seen. In order to use the side-scan sonar, all divers need to exit the water so that they do not interfere with the signals. The sonar is towed behind the boat a few feet above the bottom and an image of the acoustic readings is transmitted onto a screen in the boat. The operators will run a grid pattern in an effort to locate the victim.

Swift water rescue is another specialized area. Swift water teams rescue individuals who become trapped by raging or fast-rising waters in rivers, by dams, in storm water channels, or in raging flood waters. Initiating a successful rescue requires highly

Left: Here is a more complete view of a diver in a dry suit. Colder water temperatures will dictate the diver's use of a dry suit versus a wetsuit. This diver just left the decontamination showers as others in the background can be seen getting help rinsing off.

Above: Seven divers walk this rope line through shallow water looking for a drowning victim. The water is murky, which prevents them from seeing beyond 6 inches, so all they can hope to do at this juncture is stumble across the victim if he is lying on the bottom. The divers will continue back and forth from shore to shore covering as much area as possible while other teams of divers perform similar searches.

Left: Support personnel or tenders hold onto lines that are secured to a diver at the other end. The tenders keep track of the divers while they are in the water and communicate back and forth via signals by tugging on the rope.

Side-scan sonar involves towing a sensor behind the boat a few feet above the bottom. The acoustic readings are transmitted onto a screen in the boat where the operator looks for shape irregularities on the bottom. The operator will run a grid pattern in an effort to locate the victim, and the divers need to be out of the water. Here, the boat is being pulled slowly by rope while the operator studies the monitor.

trained personnel with ropes and sometimes inflatable boats to safely retrieve the victims without endangering the rescuers. They have to combat the pressure of the water, which is a difficult task. Swift water rescue personnel are divided into different classifications based on their training. There are shore-based support personnel, contact-rescue personnel, and those with additional training in technical rigging or the ability to assume the role of rescue sector officer. Everyone involved in the swift water rescue will be outfitted with a life jacket, and those who work near the water will have water sport–type helmets and may wear either a wetsuit or dry suit depending on the conditions of the water. The rescue might involve entering the water with a safety rope, traveling along a rope that spans the water and is secured on both sides, using an inflatable boat, throwing a rope or life preserver to the victim, or trying to reach the victim with the assistance of a helicopter.

In cold climates, accidents occur when people break through ice and fall into the water. This presents dangers to the rescuers, who do not want to plunge through ice into dangerously frigid water. The severely cold temperatures reduce the time that a victim can remain in the water before hypothermia sets in and they lose consciousness. To perform a rescue, firefighters may use airboats, sleds, ropes, dry suits, and cold-water immersion suits, which are affectionately called Gumby suits. These suits are designed for working on ice or on the surface of very cold bodies of water and are completely waterproof, allowing the diver to float with his or her head out of the water. The immediate need for medical attention to revive or warm the victim becomes critical. An ice rescue requires a rapid and efficient use of the equipment and personnel in the hopes of effecting a quick and successful rescue.

HIGH-ANGLE RESCUE

High-angle rescue teams work to free or save individuals who become trapped, stranded, injured, tangled, or scared, generally beyond the reach of ladders or aerial trucks. Rescuers work with ropes to gain access to and egress from the rescue location, and to prevent the victim from falling. They work to release the victim from whatever is trapping them, or to gain the victim's confidence in order to secure them with a harness and ropes before lowering them to the ground or to the safety of an aerial device. These rescues may involve water towers, scaffolding, overhead pipe racks, vessels or units at industrial facilities, cranes, and ledges. The equipment that this team

This decontamination (decon) unit has interior and exterior showers to decon at a hazmat incident, a dive, or other incident at the discretion of the IC. The decon unit carries no water, so a line from an engine is required. Divers will rinse off upon exiting the water.

The apparatus in this photo is used as a bomb-proof anchor supporting the bipod and lifeline being used to haul and lower the rescuers and basket litter. One firefighter is controlling the front-bumper-mounted winch to raise or lower the cable. The mini pumper is equipped with rescue equipment, medical supplies, a small water tank, and a small 250-gpm Pneumax pump. It has off-road capabilities and an extended cab to carry a crew of four, if necessary. The angled rear design permits travel over rugged and hilly terrain without damage to the rear body of the rig. A paramedic has prepared his supplies to initiate medical treatment of an injured victim.

Chapter Six

The Bronto Sky-Lift is a Finnish aerial device that is owned by the parent of E-ONE. This extremely versatile aerial device combines an elevating tower ladder design with an articulating fly section that allows for up-and-over maneuvers. Ideal for high-angle and below-grade rescue applications, the flexibility of this device is illustrated here as the fire department works in a quarry with jagged outcroppings of rock. The high-angle team was able to access the injured worker on the cliff and package the patient in a stokes basket. As rescuers lower the basket litter, firefighters in the aerial platform prepare to receive the victim and complete the evacuation. Firefighters were able to position the aerial so that they could receive the patient and subsequently strap the stokes basket into a special mounting bracket on the top of the platform for a safe journey to the ground. The coordinated use of multiple personnel and resources under a single IC decreases the risk to emergency personnel and victims while improving the odds of success.

uses centers around several types of rope, pulleys, carabiners, harnesses, stokes baskets or stretchers, and all manner of accessories and tools associated with climbing and rappelling. This work requires highly trained firefighters who are proficient in technical rescue skills involving ropes, anchoring and belaying systems, lowering and hauling systems, and stretcher work for the safe execution of the rescue.

High-angle rescuers also perform rope rescues in areas where cliffs, mountains, or other steep terrain present challenging scenarios for retrieving trapped or stranded victims.

BELOW-GRADE/TRENCH RESCUE

People who work underground can place themselves in serious danger if they do not exercise all of the proper safety precautions to ensure a secure workplace. Trench diggers, for example, risk being buried if the walls collapse while they are working in the trench. This is why OSHA regulations require the use of a steel trench box or sturdy shoring to reinforce the area that has been dug out prior to going down in the hole. The sturdiness of any trench or the amount of time

Above: Technicians place ground pads using regular plywood around the trench. These ground pads distribute the weight of the technicians, reducing the probability of a secondary collapse. Next, trench panels, 4x8-foot sheets of 14-ply plywood with 2x10- or 12-inch wood uprights bolted to the plywood, are inserted, and the placement of struts begins.

Right: With struts against uprights, additional struts are placed against the horizontal 6x6-inch timber known as a Waler. Walers are used to secure large openings needed for either rescue work or when the trench configuration does not lend itself to simply using uprights alone. The yellow flexible duct seen in this and other photos is attached to a fan. The fan blows fresh air into the trench for both the victim and rescuers. The air from the fan can be heated. Even on 90-plus-degree days, soil temperatures below grade will remain around 50 degrees Fahrenheit. This, coupled with the natural moisture of the soil, can lead to the victim developing hypothermia. Note the 5-gallon bucket used to haul the dug-out soil.

the walls can support the surrounding soil is dependent upon several conditions including the weather, the soil type and content, the moisture in the soil, and whether or not the area has been dug up previously.

Trench collapses can kill even a worker who is not totally buried. Workers are susceptible to crush injuries and asphyxiation due to the pressure of the soil against the body. Similar to the structural

building collapse scenario, rescue workers must ensure their own safety while performing the rescue mission so that they, too, do not become victims. The tools and equipment used in a trench rescue can vary from shovels, buckets, heavy timber, hammers, nails, and shoring panels to a multitude of mechanical, hydraulic, and pneumatic shoring to support the sides of a trench. Firefighters may also pump fresh air

The Tallahassee, Florida, fire department assigned this 100-foot Bronto Sky-Lift from E-ONE to their rescue house. These firefighters are skilled in all manners of technical rescue and train regularly to use the unique attributes that the Bronto has to offer. Here, they demonstrate removing a victim who was injured under a bridge. Rough terrain presents many challenges. In this case, the articulating platform allows firefighters to reach and evacuate the victim in a safe manner with minimal personnel. The Bronto allows them access that they could not achieve with any other aerial in the industry. A firefighter accompanies the victim to help keep the victim calm while another firefighter keeps them from swinging side-to-side. Carrying the victim up the steep, jagged embankment would be difficult without jostling the patient and risking further injuries.

into the trench for the victim. Some of the digging close to the victim is performed slowly by hand to prevent further injury. Typical trench rescues average 4 to 10 hours, although many last in excess of 12 hours. Victims completely buried for more than 1 hour often succumb to their injuries. In these cases, rescuers often switch to a recovery mode where more cautious and deliberate actions are used to reduce the possibility of injury or death to themselves.

Scene safety is important in any rescue situation, but trench collapses are especially dangerous. To understand the significance of the hazard, consider that the weight of one cubic yard of soil is close to 3,000 pounds, which corresponds to the weight of a mid-size automobile. When the walls of a trench collapse, the volume of soil falling on a person can involve upwards of 3 to 4 cubic feet. Untrained personnel attempting to rescue a trapped worker can themselves become victims. There are reports that as

many as 65 percent of all would-be rescuers become trapped and suffer traumatic injuries or death. Due to the soil's weight and pressure, victims must be completely dug out to the soles of their feet. Any attempt to pull a partially buried victim out using a rope is futile and may result in serious internal injuries or dismemberment.

HAZARDOUS MATERIALS

Chemical use in our society has grown in the past decades to support significant advancements in many phases of our lives, but consequently, this represents a multitude of potential dangers to the general public. Whether the dangers involve a liquid or powder spill, a toxic release of vapors, an explosion, or any number of hazardous situations, the fire departments remain the first line of defense to mitigate the situation. These events can be generated from business *continued on page 158*

Left: Chicago firefighters survey the wreckage after a building collapse on the west side of the city. Contractors working to demolish the neighboring structure breached the integrity of a common wall, and the roof collapsed. Fortunately, there were no injuries. Saws, a water extinguisher, and other tools lie near the firefighters in the event that they have to work to free a victim.

Below: This enormous US&R rig belongs to the fire department in Alhambra, California. This is a first-of-its-kind rig built by American LaFrance as a tractor and trailer combination. The lumber stored in the rear is for below-grade rescue work. This rig carries a full complement of rescue tools, including hydraulic spreaders, cutters, rams, air bags, torches, saws, hazmat meters, fans, cribbing, and supplemental lighting. Access to the top compartments is via a ladder above the lumber, which hinges to the ground to create a stairway.

Above: A one-of-a-kind rig is pictured here. The City of Orange Fire Department (ORG) in Southern California has for many years maintained a fleet of Seagrave fire apparatus. It seemed only natural, then, when the department needed to design a combination heavy rescue and US&R rig that it would involve Seagrave. Although Seagrave did not build this type of unit at the time, ORG went to Super Vac for the rescue body and Seagrave for the cab and chassis. Now this is certainly not the only tractor-trailer combination for a US&R rig. What is unique is that this is the only tractor-drawn US&R with a tiller! The department purchased a tractor and tillered aerial trailer, minus the body and aerial device. Without the aerial substructure there was a tremendous amount of space available for equipment storage, and they maintained the maneuverability of this very long rig by utilizing the rear steering tiller.

Right: Four-by-four wood timbers and screw jacks are used as struts to secure the trench walls. Since the plywood sheets only hold back the loose dirt, it is the 2x12-inch upright that actually carries the load. Note the wood blocks and wedges behind the plywood. These fill voids and ensure the strut and upright completely secure the trench wall from secondary collapse.

The Chicago Fire Department Hazardous Incident Team (HIT) 5-1-1 speeds by en route to a call. The rig was built by American LaFrance on an Eagle chassis and is identical to 5-1-2 used by the south-side HIT. The HIT is supported by specially trained engine and truck companies throughout the city that will also respond to incidents involving hazardous materials.

Firefighters have secured this car with cribbing. A paramedic is inside with the victim as another firefighter uses a hydraulic cutter to separate the car's roof. It's hard to see, but the victim and paramedic are protected from flying debris because they are under a tarp inside the car. After the roof is pulled off the car, the firefighters will be able to move the victim onto a backboard.

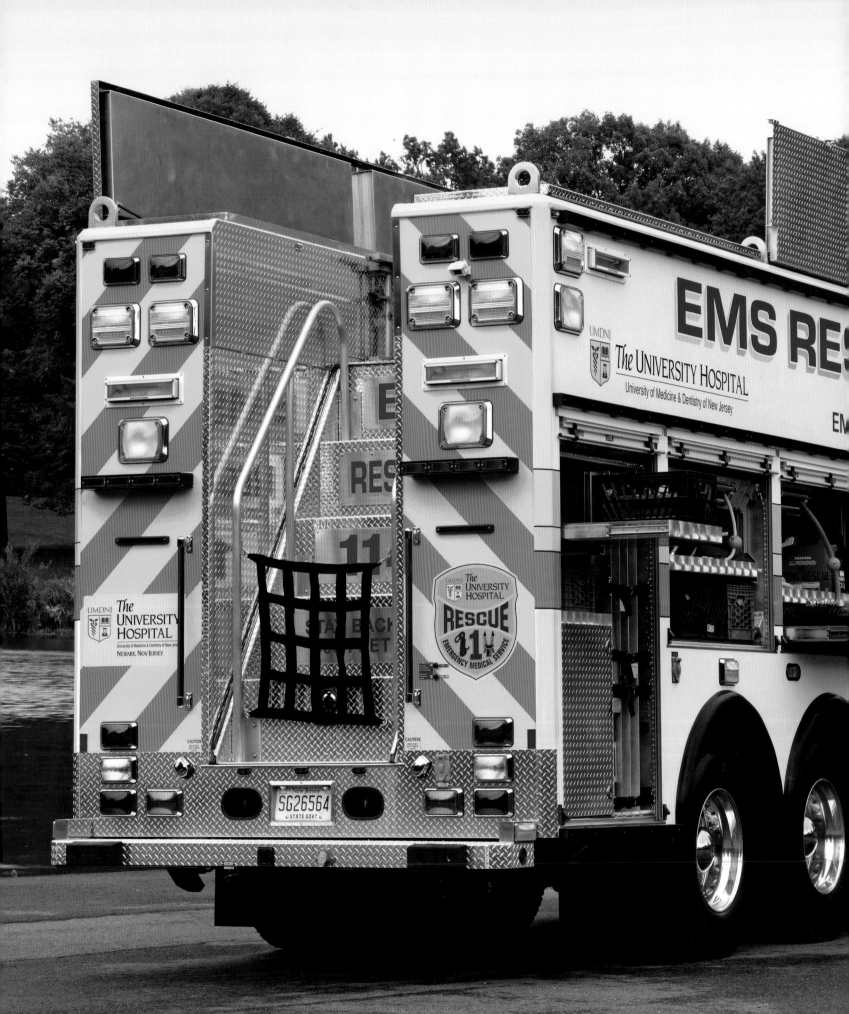

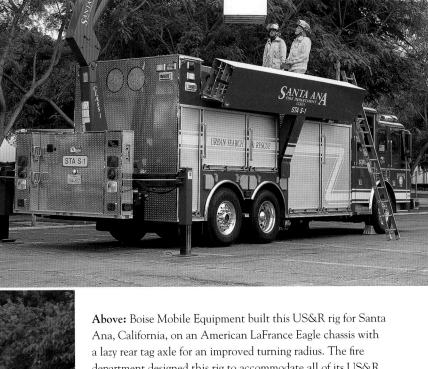

Above: Boise Mobile Equipment built this US&R rig for Santa Ana, California, on an American LaFrance Eagle chassis with a lazy rear tag axle for an improved turning radius. The fire department designed this rig to accommodate all of its US&R supplies, in addition to the heavy rescue tools that its members utilize more often. The department incorporated a knuckle boom crane on the back of the rig to access storage pods that are neatly tucked out of the way of the tools that are used more often. At a US&R deployment, the firefighters deploy a ground ladder from the hydraulic ladder rack to climb on top of the truck. There, by remote control, they operate the crane to grab one of the pods that is only accessible from the top.

Left: The University of Medicine and Dentistry, a state-run institution with several campuses in New Jersey, runs the EMS and heavy rescue for the City of Newark, the only hospital-based heavy rescue in the United States. In 2006, the university put this American LaFrance heavy rescue in service. It staffs the unit with two rescue specialists and responds to vehicle extrications, high-angle and below-grade rescues, water rescues, and confined-space incidents. Some equipment is visible on this side of the rig. The rear staircase can be raised hydraulically to allow access to lumber and shoring materials that are stored down the center of the body.

The Deptford Township Fire Department in Gloucester County, New Jersey, runs this big heavy rescue. Built by Super Vac on a Spartan Gladiator chassis, this unit is staffed by career firefighters during the daytime and volunteers at night and on the weekends. A close view with the compartments open shows a portion of the tools and equipment on board. Individual slots and slide-out trays maximize the fire department's ability to pack as much into the rig as possible so that it can meet any task that it is called upon to handle. At the rear is a stairway to the top of the rig where there is even more storage.

continued from page 152

facilities that use hazardous materials, during transport of hazardous materials, or by illegal activities like clandestine drug laboratories or illegal dumping. There are over a million known hazardous chemicals, 70,000 of which are regulated and transported daily in the United States. With the added threat of weapons of mass destruction (WMD) and acts of terrorism, current hazmat personnel are trained in the detection and mitigation of WMD, as well as chemical warfare agents. They are equipped with meters and detection equipment for nerve agents, blister agents, biohazards, and radiological isotopes.

Hazardous materials response is a discipline that is mastered through information management. The initial actions at an incident must be taken quickly but with careful consideration to how the chemical will react under the release conditions that occur at the scene. This assessment requires accurate identification, appropriate classification, adequate understanding of physical and chemical properties of the substance, and methods for containment and disposal.

This is another area of specialized training, tools, equipment, and vehicles for the fire service. Hazmat operations include the identification of the substance and its containment and cleanup. Portions of a hazmat scenario may be carried out by private contractors who specialize in this type of work. Among the gear required for the hazmat team are various types of protective suits to safeguard firefighters who move into the hot zone, which refers to the area of a scene where the toxic substances are most dangerous. There are four levels of protective clothing. The highest level of protection is level A, which is total encapsulation of the firefighter, including the SCBA. This protects against vapors, gases, liquids, mists, and particles. Next is level B, which is to be worn when vapor protection is not required. Here, in addition to SCBA, protective suits will include gloves, boots, face pieces, and hoods, which are secured to prevent the splashing of liquids. Level C protection is similar to level B. It does not require the firefighter to use SCBA but does specify the use of some form of air-purifying respirator. The lowest level is full structural PPE.

EPILOGUE

Fighting fire is and has always been a noble profession. Those who serve the public, whether by choosing the fire service as a career or as a volunteer, share a sense of duty and demonstrate unparalleled bravery. These professionals are dedicated, motivated, and driven by this sense of duty to be the first responders to incidents of multiple dimensions.

The job of firefighters today is continually changing. The amount of training and knowledge required to perform this job has never been greater due to the dynamic nature of the incidents and events that they are called upon to handle. The specialized equipment alone that a firefighter uses is more sophisticated and complex than it ever has been.

Going forward, firefighting will continue to evolve as the profession is faced with even greater diversity and new challenges. Firefighters will expand their expertise to meet these challenges because they are our first line of defense in a world fraught with the ever-present dangers of fire, medical emergencies, accidents, natural disasters, and terrorism. Firefighting is a career that is attractive not only because of its noble attributes, but because there are a multitude of opportunities within the fire service that can be explored. Urban, suburban, hazmat, industrial, and wildland firefighting, plus rescue and EMS responses offer broad opportunities for professional growth.

INDEX